FATIGUE of MATERIALS III
ADVANCES and EMERGENCES in UNDERSTANDING

Proceedings of the third biennial symposium

Cover Photograph: Scanning electron micrographs of the longitudinal test sample of 300 M alloy steel cyclically deformed under stress control at a maximum stress of 369 MPa and a fatigue life of 166,549 cycles, showing coplanar array of fine microscopic cracks in the region prior to onset of unstable crack growth.

(from K. Manigandan, T.S. Srivatsan, G. Doll, and T. Quick)

FATIGUE of MATERIALS III
ADVANCES and EMERGENCES in UNDERSTANDING

Proceedings of the third biennial symposium

Sponsored by

The Minerals, Metals & Materials Society (TMS)
and ASM International

held during

Materials Science & Technology 2014 (MS&T'14)
October 12–16, 2014 in Pittsburgh, Pennsylvania

MS&T14®

Edited by

T.S. Srivatsan M. Ashraf Imam

R. Srinivasan

Editors
T.S. Srivatsan
M. Ashraf Imam
R. Srinivasan

ISBN 978-3-319-48597-3 ISBN 978-3-319-48240-8 (eBook)
DOI 10.1007/978-3-319-48240-8

Chemistry and Materials Science: Professional

Published by Springer International Publishers, Switzerland, 2016
Reprint of the original edition published by John Wiley & Sons, Inc., 2014, 978-1-119-04148-1

Printed on acid-free paper

This Springer imprint is published by Springer Nature
The registered company is Springer International Publishing AG
The registered company address is: Gewerbestrasse 11, 6330 Cham, Switzerland

TABLE OF CONTENTS
Fatigue of Materials III: Advances and Emergences in Understanding

Aluminum

Advanced Materials

Composites

Ferrous Materials

Modeling

PREFACE

This volume contains the papers presented in the third international symposium titled *Fatigue of Materials III: Advances and Emergences in Understanding* held during the Materials Science and Technology 2014 meeting at Pittsburgh, Pennsylvania, USA in October 2014. The six-session symposium was sponsored by the Mechanical Behavior of Materials Committee of The Minerals, Metals & Materials Society (TMS) and ASM International. It is essentially the seventh in a series of symposia on the topic of fatigue of materials. The first three symposia focused on deformation and fracture of ordered intermetallic materials: the first symposium was held in Pittsburgh in 1993; the second symposium was held in Rosemont (Illinois, USA) in 1994, and the third symposium was held in Cincinnati (Ohio, USA) in 1996. The fourth symposium was in honor of Professor Paul C. Paris and focused on high cycle fatigue of structural materials and held in Indianapolis (Indiana, USA) in 1997. The fifth symposium related to advances and emergences in understanding was held in Houston (Texas, USA) in October 2010. The sixth symposium also related to advances and emergences in understanding was held in Pittsburgh in October 2012.

This symposium was well represented with abstracts from engineers, technologists, and scientists from academia, research laboratories, and industries, located both within the United States and few countries overseas. The 30-plus abstracts that were approved for presentation at the symposium were divided into five sessions:

Session 1:	Aluminum Alloys
Session 2:	Ferrous Materials I
Session 3:	Ferrous Materials II
Session 4:	Composite Materials
Session 5:	Advanced Materials
Session 6:	Modeling

The abstracts chosen for presentation at the symposium cover a broad spectrum of topics that represent the truly diverse nature of the subject of fatigue as it relates to the world of materials. The domain and importance of materials has grown appreciably in strength and significantly in stature to become a key area of scientific and applied research. We, the co-organizers, have made every attempt to bring together individuals who could in a positive way put forth the advances while concurrently striving to enhance our prevailing understanding of the topic of fatigue of materials. We extend our warmest thanks and appreciation to both the authors and session chairmen for their enthusiastic commitment and participation.

We also extend our most sincere thanks and appreciation to elected representatives that serve on the Mechanical Behavior of Materials Committee of ASM International and TMS for their understanding and acknowledgment of our interest and approval of our request to organize this intellectually stimulating event. An overdose of special thanks, gratitude, and appreciation are reserved and extended to **Ms. Trudi Dunlap** (Programming Manager at TMS) for her sustained attention, assistance, interest, involvement, and participation stemming from understanding and enthusiastic willingness to help. This ensured a timely execution of the numerous intricacies related to smooth orchestration and layout of this symposium from the moment immediately following its approval and up until compilation and publication of this publication. At moments of need, we the symposium organizers have found her presence in TMS and participation to be a pillar of support, courteous, understanding, professional, and almost always enthusiastically helpful and receptive. Timely thanks and assured appreciation is also extended to **Mr. Robert (Bob) Demmler** (Graphic Designer, TMS) for his efforts and enthusiasm in ensuring organization and compilation of the material included in this bound volume in a cohesive, convincing, and compelling manner. Special thanks and appreciation is also extended to **Mr. K. Manigandan** (University of Akron) for his time and efforts in formatting few of the submitted manuscripts very much in conformance with instructions for manuscript preparation for inclusion in this volume. The timely compilation and publication of this volume would not have been possible without the cooperation of the authors and the publishing staff at TMS (Warrendale, PA, USA).

We truly hope that this volume will provide engineers, scientists, and technologists with new perspectives and directions in their research endeavors toward evaluating, understanding, and improving the fatigue behavior of materials, spanning the entire spectrum to include both engineering materials and engineered materials.

Dr. T. S. Srivatsan
The University of Akron
E-Mail: tsrivatsan@uakron.edu

Dr. M. Ashraf Imam
George Washington University
E-Mail: ashrafimam@gwu.edu

Dr. R. Srinivasan
Wright State University
E-Mail: raghavan.srinivasan@wright.edu

SYMPOSIUM ORGANIZERS

Dr. T.S. Srivatsan is Professor of Materials Science and Engineering in the Department of Mechanical Engineering at The University of Akron. He received his graduate degrees [Master of Science in Aerospace Engineering (M.S. 1981) and Doctor of Philosophy in Mechanical Engineering (Ph.D. 1984)] from Georgia Institute of Technology. Dr. Srivatsan joined the faculty in The Department of Mechanical Engineering at The University of Akron in August 1987. Since joining, he has instructed undergraduate and graduate courses in the areas of Advanced Materials and Manufacturing Processes, Mechanical Behavior of Materials, Fatigue of Engineering Materials and Structures, Fracture Mechanics, Introduction to Materials Science and Engineering, Mechanical Measurements, Design of Mechanical Systems and Mechanical Engineering Laboratory. His research areas currently span the fatigue and fracture behavior of advanced materials to include monolithic(s), intermetallic, nano-materials, and metal-matrix composites; processing techniques for advanced materials and nanostructure materials; inter-relationship between processing and mechanical behavior; electron microscopy; failure analysis; and mechanical design. His funding comes primarily from both industries and government and is of the order of a few millions of dollars since 1987. A synergism of his efforts has helped in many ways to advancing the science, engineering and technological applications of materials.

Dr. Srivatsan has authored/edited/co-edited fifty-three (53) books in areas cross-pollinating mechanical design; processing and fabrication of advanced materials; deformation, fatigue, and fracture of ordered intermetallic materials; machining of composites; failure analysis; and technology of rapid solidification processing of materials. He serves as co-editor of *International Journal on Materials and Manufacturing Processes* and is on the editorial advisory board of journals in the domain of Materials Science and Engineering. His research has enabled him to deliver over two-hundred plus (200+) technical presentations in national and international meetings and symposia; technical/professional societies; and research and educational institutions. He has authored and co-authored over six-hundred and fifty plus (650+) archival publications in international journals, chapters in books, proceedings of national and international conferences, reviews of books, and technical reports. In recognition of his efforts andcontributions and his impact on furthering science, technology, and education he has been elected Fellow of American Society for Materials, International (ASM Int.); Fellow of American Society of Mechanical Engineers (ASME); and Fellow of American Association for the Advancement of Science (AAAS). He has also been recognized as Outstanding Young Alumnus of Georgia Institute of Technology, and outstanding Research Faculty, the College of Engineering at The University of Akron. He offers his knowledge in research services to the U.S. Government (U.S. Air Force and U.S. Navy), National Research Laboratories, and industries related to aerospace, automotive, power-generation, leisure-related products, and applied medical sciences . He has the distinct honor of being chosen, in recent years, for inclusion in *Who's Who in American Education, Who's Who in the Midwest, Who's Who in Technology, Who's Who in the World, Who's Who in America, Who's Who in Science and Engineering, Who's Who among America's Teachers,* and *Who's Who among Executives and Professionals* (Cambridge).

Dr. M. Ashraf Imam currently holds the position of Professor of Materials Science at George Washington University in Washington, DC. Dr. Imam obtained a D.Sc. degree in Materials Science from George Washington University and an M.S. from Carnegie-Mellon University in the field of Metallurgy and Materials Science. In the immediate past he was Research Metallurgist at the Naval Research Laboratory (NRL) and pursued basic research on aspects related and relevant to material structure-property relationships in materials. Dr. Imam has demonstrated a unique ability to analyze the technical problems and design various experimental and theoretical techniques to arrive at optimum solutions. Often these investigations have resulted in technological breakthroughs leading to useful and unique applications. His publications (over 200) in a variety of refereed journals ranging from *Acta Metallurgica* to *Philosophical Magazine* are a clear testimony of Dr. Imam as a researcher and teacher with outstanding creativity and resourcefulness in the area of Materials Science and Engineering. His papers cover a variety of topics with a unifying theme of microstructure, characterization, and analysis. His expertise on titanium resulted in editing three books on titanium including a large chapter on titanium in *Kirk-Othmer Encyclopedia of Chemical Technology*. These contributions to the literature have been widely acclaimed and recognized. Dr. Imam is Fellow of American Society for Materials International (ASM Int.). He is also a recipient of the George Kimbell Burgess Award. The award is given in recognition of outstanding achievement in research or administration to a member of the ASM Washington, DC chapter who has made original contributions in the field of metallurgy, materials, or mechanics or who demonstrates outstanding leadership in those fields.

Dr. Raghavan Srinivasan, Professor in the Mechanical and Materials Engineering Department, is the Director of the Materials Science and Engineering Program at Wright State University, Dayton, Ohio. He received his Ph.D. from the State University of New York, Stony Brook, and his M.S. from the University of Florida, Gainesville, both in Materials Science and Engineering, after completing his bachelors degree from the Indian Institute of Technology, Madras. He has over 100 technical publications in the areas of processing and mechanical behavior of advanced materials. He has edited two books, and contributed to the ASM Handbook series. His work has resulted in four patents and three patent applications which are under review. Dr. Srinivasan is a member of Tau Beta Pi, Alpha Sigma Mu, and Phi Kappa Phi honor societies, and has been recognized as Fellow of Alpha Sigma Mu, the International Honor Society for Materials Science and Engineering. He is a member of TMS, ASM International, and SAMPE. He serves as an evaluator for ABET, and is a member of several committees, including the Accreditation, Professional Registration, Mechanical Behavior, and Shaping and Forming Committees of TMS, and is an associate member of the Failure Analysis Committee of ASM International. Dr. Srinivasan is a registered Professional Engineer in the State of Ohio.

SESSION CHAIRS

Dr. Ly L. Alexander
Colorado School of Mines, USA

Dr. R. Ebara
Fukuoka University, Japan

Dr. Mikado Hiroko
YKK Corporation, Japan

Dr. Sushant Jha
Air Force Research Laboratory, USA

Dr. Masuda Kenichi
Toyama University, Japan

Dr. Yong-Nam Kwon
Korea Institute of Materials Science, Korea

Dr. Roy J. Matthew
University of Manchester, United Kingdom

Dr. K. S. Ravi Chandran
University of Utah, USA

Dr. G. Sasikala
IGCAR-Kalpakkam, India

Dr. Zihai Shi
Nippon Koel Company, Ltd., Japan

Dr. Jesus Toribio
University of Salamanca, Spain

FATIGUE of MATERIALS III
ADVANCES and EMERGENCES in UNDERSTANDING

Aluminum

Fatigue of Materials III: Advances and Emergences in Understanding
Edited by: T.S. Srivatsan, M. Ashraf Imam, and R. Srinivasan
TMS (The Minerals, Metals & Materials Society), 2014

IMPROVEMENT OF FATIGUE PROPERTIES OF CAST ALUMINUM ALLOY A356 BY WARM DEFORMATION

M. J. Roy[1], D. M. Maijer[2], Y. Nadot[3]

[1] **The University of Manchester**
School of Mechanical, Aerospace and Civil Engineering
Sackville Street Building, Manchester
United Kingdom, M13 9PL

[2] **The University of British Columbia**
Department of Materials Engineering
309-6350 Stores Road, Vancouver, BC
Canada, V6T 1Z4

[3]Institut PPRIME - CNRS
Université de Poitiers
ENSMA - UPR 3346 -
Départment Mécanique des Matériaux
Téléport 2 - 1 Avenue Clément Ader - BP
4019 - 86961 FUTUROSCOPE CHASSENEUIL
Cedex - **France**

Abstract

Aluminum foundry alloys such as A356 are used extensively in applications where high cycle fatigue (HCF) resilience is a key design consideration. Fully reversed, multiaxial HCF studies on this alloy in the T6 condition have shown that endurance limits are governed by maximum principal stress and driven by crack propagation as opposed to initiation. In light of these fatigue characteristics, it has been found that warm deformation imparted via flow forming prior to heat treatment improves the fatigue resilience by upwards of 30% depending on the degree of deformation. The fatigue performance improvement has been attributed to eutectic particle size and morphology changes and potential recrystallisation of the primary phases. This hypothesis is supported by extensive particle characterisation and preliminary EBSD results.

Keywords: A356, Fatigue, Flow forming

1. Introduction

Aluminum castings are employed by many different industries to provide lightweight, near net-shape components. However, these components may be defective due to inherent porosity and microstructure inhomogeneity. Standard forging operations can mitigate some of these issues, but are expensive in terms of tooling and operation costs. A lower cost and more flexible alternative for hollow axisymmetric components is rotary forming. This process consists of fixing the workpiece on a rotating mandrel followed by local deformation with a roller or other shaping tool which moved axially along the component. This introduces a highly localized plastic zone which reduces the wall thickness and elongates the component. Rotary forming is often commercially referred to as spinning, shear forming or flow forming depending on the degree of deformation imparted to the workpiece [1].

A356 (Al-7Si-0.3Mg) is a common aluminum casting alloy consisting of a dendritic structure comprised of α-Al surrounded by Si-rich eutectic (Fig. 1) which typically undergoes a T6 heat treatment after casting. High cycle fatigue (HCF) lives within this material are typically decided by porosity and large eutectic particles or other inclusions remaining post-heat treatment [2]. Previous work [3, 4] which investigated the multiaxial fatigue properties of this alloy have indicated that fatigue life limiting defects are quite large, upwards of 400 μm as crack propagation dominates HCF behaviour. Outside of pure torsion, the maximum principal stress is the primary driver of crack propagation which indicates that S-N data tabulated from uniaxial fatigue testing may be employed to gauge multiaxial performance.

2. Experiment details

Two sources of flow formed A356 were used to characterize the effects on fatigue resilience due to processing. Material that was commercially formed and heat treated to the T6 condition was received as unmachined, low pressure die cast wheel blanks from a North American automotive wheel manufacturer, referred to further as `commercial'. The second source of material was modified as-cast (AC) wheel blanks from the same supplier. A portion of these blanks were T6 treated immediately and the remainder processed with a purpose-built experimental rotary forming apparatus, referred to further as `experimental'. In both cases, the material was deformed at a target temperature of 350°C, with the commercial material experiencing heavy deformation (upwards of 50% wall thickness reduction) and the experimental material undergoing light deformation (approximately 10% wall thickness reduction). All material in this study was subjected to a standard T6 heat treatment consisting of solutionizing at 540°C for 3 hours, followed by a 60°C water quench and concluded with an artificial aging period at 150°C for 3 hours immediately after quenching. The effect of deformation on the microstructure showed either a light deformation of the dendrite cells in the direction of forming or a complete loss of dendritic structure. The former microstructure was predominant in the experimental material, while the commercial material exhibited a range of microstructure between the two conditions.

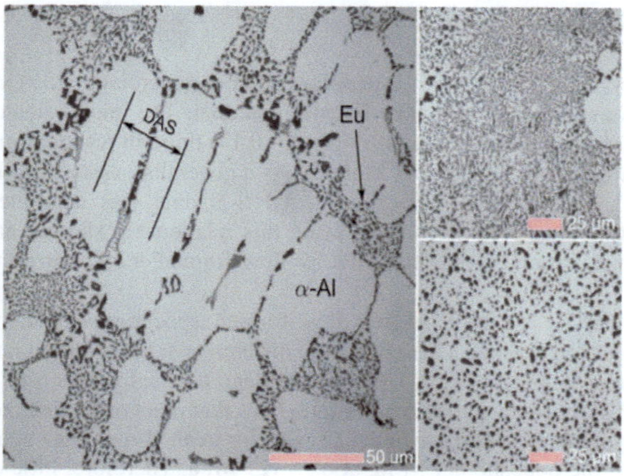

Figure 1. As-cast A356 microstructure (left), as-cast eutectic (top right) and eutectic post T6-treatment (bottom right).

Table 1. Source, condition and quantity employed in fatigue study.

Source	Condition	Quantity
Comm.	AC	55
	Formed	4 (A) 24 (AD) 4 (H) 4 (HD)
Exp.	AC	15
	Formed	12

Post heat-treatment, hour-glass uniaxial fatigue specimens were extracted with geometry conforming to ASTM E606. The commerically-formed material had specimens extracted from different locations within the blank, corresponding to heavy and light deformation in the axial (A and AD, respectively) and hoop direction (H and HD, respectively). Samples from the experimentally-formed material were drawn solely in the axial direction from the most heavily deformed region of the component. As cast, undeformed material (AC) corresponding to both blank configurations had specimens extracted from equivalent locations. See Table 1 for specimen counts, and Fig. 2 for specimen extraction points.

All specimens were run-out tested to failure under fully reversed conditions (R = -1) with a Sonntag eccentric fatigue machine at 33 Hz, with load amplitudes targeting the mid to high cycle (10^5-10^6) fatigue regime. The microstructure was analyzed to ascertain effects of forming using microindentation, optical and scanning electron microscopy (OM and SEM) techniques.

3. Fatigue Results

Fig. 3 demonstrates S-N data for undeformed (I commercial and II experimental) versus flowformed (III commercial and IV experimental) material. The improvement in endurance has been quantified by fitting Basquin relationships (b = 0.17) through results from each sample type and source via non-linear least-squares (Table 2). As a general assessment of how the endurance limit improved, the commercial and experimental S-N data was combined and fitted. The endurance limit(sf) evaluated at 106 cycles using this relationship showed an provement of 75 MPa for undeformed material versus 107 MPa for flow-formed or an improvement of 30%.

Separate analysis on each of the flow-formed specimen streams (experimental vs. commercial) was also conducted. There was little difference in fatigue limit noted for each of the commercial specimen orientations (AD vs. HD) or degree of deformation (A vs. AD and H vs. HD). At least four of the A specimens had fatigue lives that meet or exceed the AD specimens at the same load amplitude. One HD specimen failed prior to an H specimen at a load amplitude of σ^a = 110 MPa with the opposite occurring at σ^a = 125 MPa. Therefore, little conclusions can be drawn regarding the effect of orientation and deformation within the scatter of S-N data. Overall, the commercial material showed a 33% increase in fatigue limit over undeformed material. The experimental material showed approximately the same fatigue lives as the specimens taken from the commercial material for loading amplitudes between 110 and 120 MPa, and other specimens had lives falling within two orders of magnitude of the commercial specimens with identical σ^a. The undeformed experimental material demonstrated significantly higher fatigue limits than the undeformed commercial material (72 vs. 84 MPa, respectively), which accounts for the reason why experimental material only showed a 19% increase in fatigue limit due to forming. This is likely due to a specimen size effect; larger specimens have an increased number of defects which could affect fatigue limits.

Overall, flow-forming has been shown to significantly improve the fatigue properties of A356. Departure from fitted Basquin relationships, quantified by both R^2 and RMSE was diminished for both sources of material. As the improvement in fatigue limits for commercial material was appreciably higher than the experimental material, increased deformation and resulting porosity mitigation is the principal reason.

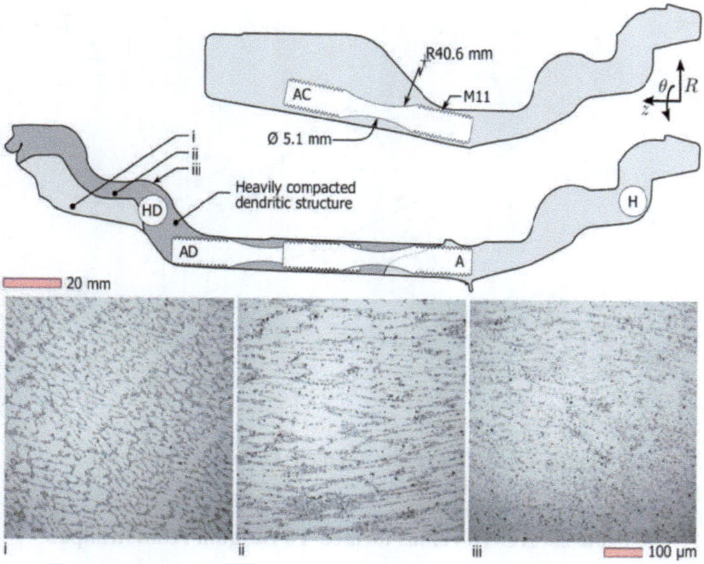

Figure 2. **Pre and post deformed geometry of commercial specimens. Micrographs correspond to locations in the diagram above**

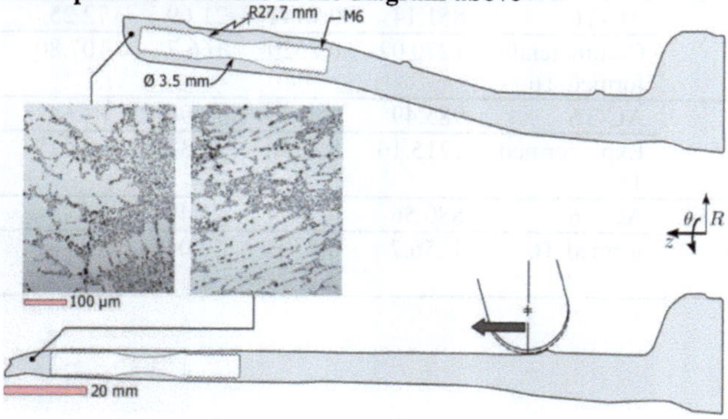

Figure 3. **Pre and post deformed geometry of experimental specimens. Micrographs correspond to locations in the diagram above.**

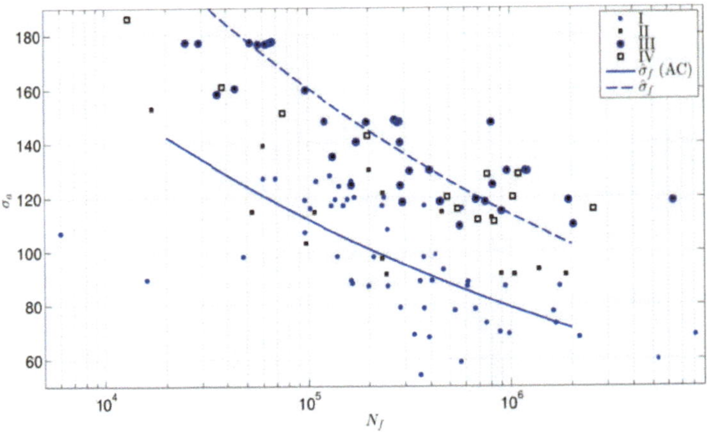

Figure 4. AC-T6 versus formed-T6 fatigue results.

Table 2. AC-T6 versus formed-T6 Basquin relationships

Fig. 3 Series	**Conditions**	**σ'$_f$ (MPa)**	**R^2**	**RMSE**	**σ_f | 10^6**
I	AC-T6	851.14	0.7441	21.69	72.25
III	Commercially formed-T6	1270.02	0.8629	166.7	107.80
II	AC-T6	985.49	0.8322	15.90	83.64
IV	Exp. formed-T6	1215.16	0.8850	17.32	103.15
I & II	AC-T6	880.56	0.7454	21.49	74.75
III & IV	formed-T6	1256.7	0.8703	17.04	106.67

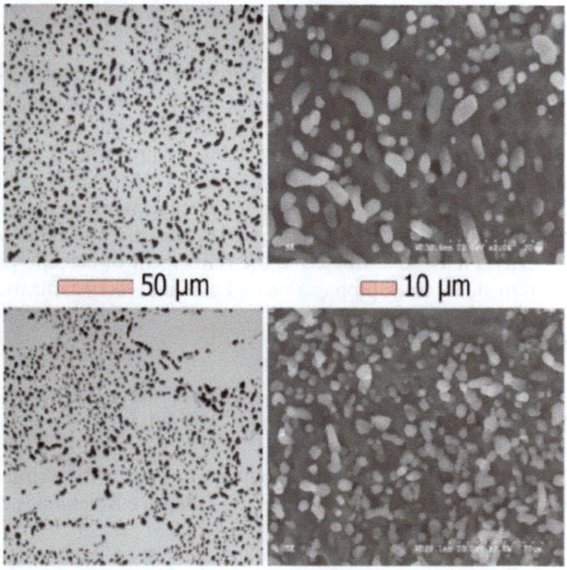

Figure 5. Undeformed A356-T6 eutectic particles (top) versus deformed (bottom)

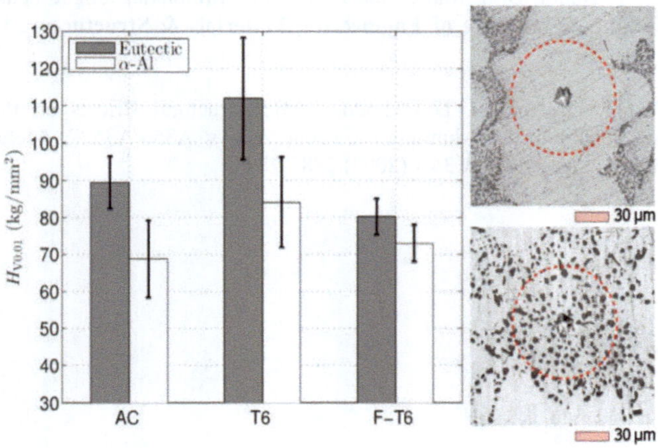

Figure 6. Phase-specific microhardness measurements of untempered (AC), undeformed-T6 (T6) and deformed-T6 (F-T6)

5. Summary

In the case of rotary formed material, a comparison of runout S-N data showed the overall HCF endurance limit was improved by 30% over undeformed material. The commercially processed material showed a 33% increase in endurance limit, and the experimentally flow formed material showed a 19% improvement. As the commercially processed material was deformed to a much higher extent than the experimentally processed material, this implies that increased levels of rotary forming improves the fatigue properties of the material accordingly. The cause for this increase is primarily attributed to the mitigation of porosity. Secondary causes may be attributed to the refinement of eutectic particles induced by processing, which in turn slows crack propagation and potential recrystallization.

6. References

1. O. Music, J.M. Allwood, K. Kawai, “A review of the mechanics of metal spinning”, **Journal of Materials Processing Technology** 210 (2010) 3-23

2. D.L. McDowell, K. Gall, M.F. Horstmeyer, J. Fan, “Microstructure-based fatigue modeling fo cast A356-T6 alloy”, **Engineering Fracture Mechanics** 70 (2003) 49-80.

3. M.J. Roy, Y. Nadot, C. Nadot-Martin, P.-G. Bardin, D.M. Maijer, “Multiaxial Kitigawa analysis of A356-T6”, **International Journal of Fatigue**, 33 (2011), 823-832.

4. M. Roy, Y. Nadot, D.M. Maijer and G. Benoit, “Multiaxial fatigue behavior of A356-T6”, **Fatigue & Fracture of Engineering Materials & Structures**, 35 (2012), 1148-1159.

5. Q.G. Wang, D. Apelian, D.A. Lados, “Microstructural Effects on the Tensile and Fracture Behaviour of Aluminum Casting Alloys A356/A357”, **Metallurgical and Materials Transactions A** 34A (2003) 2887-2899.

ROLE OF DISPERSOIDS ON THE FATIGUE BEHAVIOR OF ALUMINUM ALLOYS: A REVIEW

Raghavan Srinivasan
Department of Mechanical and Materials Engineering,
Wright State University
Dayton OH 45435, USA

M. Ashraf Imam
Department of Mechanical and Aerospace Engineering
The George Washington University
Washington, DC 20052, USA

ABSTRACT

Aluminum alloys contain dispersoids, namely thermodynamically stable second phase particles ranging in size from 0.1 to 1.0 μm, which are used, for example, to control grain size. The presence of second phase particles could have a beneficial or detrimental effect on fatigue behavior depending on their properties and distribution within the aluminum matrix. Hard non-shearable particles promote cross-slip and deformation homogenization that delays fatigue crack initiation, while shearable particles result in planar slip, which promote strain reversal and crack closure and slows crack propagation. Cracking and fragmentation of the particles, or strain incompatibility at the interface of the dispersoids with the matrix, can lead to the nucleation of cracks ahead of the main advancing crack. This paper presents a review of the effect of dispersoids on the fatigue behavior of aluminum alloys.

1. Introduction

Fatigue failure or failure of an engineering component that occurs after repetitive or fluctuating loads has been attributed to over 90% of service related failure of metallic components. Based on the stress level, fatigue behavior has been broadly classified into high cycle fatigue (HCF), when the stress is below the macroscopic yield strength of the metal, and low cycle fatigue (LCF), when the stress level exceeds the yield stress over a part of the loading cycle. LCF failures are broadly grouped into failures that occur after less than about 10^4 cycles whereas HCF failures occur after 10^6 or more cycles. There is, however, no definite boundary that separates LCF from HCF behavior, because, irrespective of the type of failure, fatigue failure typically progresses through three stages, the initiation of a crack, the growth of the crack, and finally fast fracture when the remaining cross section of the component is no longer able to support the applied load, as shown in Figure 1. The initiation and growth stages of fatigue failure involve the movement of dislocations, while the final fracture is related to the fracture toughness of the material.

Models for fatigue crack growth (FCG) that are based on linear elastic fracture mechanics relate FCG rate to the range of stress intensity factor ΔK resulting from the variation in stress. If there is a pre-existing crack of length a in a sample, and the stress varies between σ_{min} and σ_{max}, then the stress intensity factor range is $\Delta K = K_{max} - K_{min} = \sigma_{max}\sqrt{\pi a} - \sigma_{min}\sqrt{\pi a}$ In Figure 1, the FCG rate is essentially zero when ΔK is below a threshold ΔK_{th}.[1] Under these conditions, though cracks may be present in the material, they do not grow. In Region II, cracks grow as a function of ΔK. The Paris model for the growth of cracks is given by: $da/dN = A(\Delta K)^n$; where the LHS is the rate at which the crack length a increases for each cycle of fatigue loading. A and n are empirical material constants. During fatigue loading, the crack length increases continuously, as a result ΔK, also increases. Once the crack length reaches a critical value, there is a rapid and unstable crack growth because K_{max} has reached a value equal to the fracture toughness of the material K_{IC}.

Improvements in fatigue behavior can be achieved by addressing the factors that influence the initiation and growth stages of cracks. At the macroscopic level, this involves elimination mechanical stress concentrators such as notches, localized changes

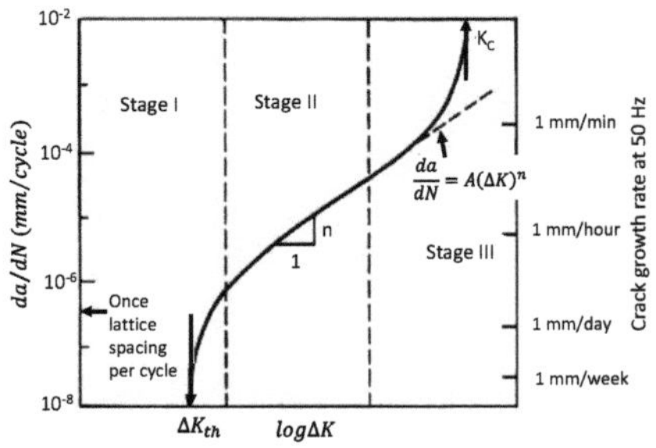

Figure 1. Fatigue crack growth (FCG) as a function of ΔK [1]

in cross section, and poor surface finish. These changes help in delaying crack initiation. At the microscopic level, factors that influence ease of dislocation motion, such as alloying, grain size, morphology of multiphase materials, and the presence of "second-phase" particles, also affect the fatigue behavior of the metal. In this paper, we look at the influence of second phase particles on the fatigue behavior of aluminum alloys.

Second phases in aluminum alloys can be classified into three main groups [2,3,4]: (a) inclusions, which are typically large particles (>5μm) of impurities from the manufacturing process, such as aluminum oxide and other compounds from the slag, (b) dispersoids, which are defined as thermodynamically stable particles ranging in size from 0.1 to 1 μm, are precipitated during thermomechanical processing (TMP), and do not dissolve at solution heat treatment temperatures associated with subsequent tempering treatments, and (c) precipitates, which are formed by controlled precipitation of hard particles through TMP, and which re-dissolve into the matrix at increased temperatures. Dispersoid particles typically contain Cr, Mn, or Zr. These alloying elements are added to control grain size of age hardenable wrought aluminum alloys (2xxx, 6xxx, 7xxx, and some 8xxx) [2]. Table 1 lists the dispersoids that are present in some aluminum alloys [5].

Table 1: Dispersoids in Aluminum Alloys

Alloy	Dispersoid	Alloy	Dispersoid
2090	Al_3Zr	6013	$Al_{12}Mn_3Si$
2091	Al_3Zr	7046	Al_3Zr
2095	Al_3Zr	7x50	Al_3Zr
2x24	$Al_{20}Cu_2Mn_3$	7055	Al_3Zr
3003	$Al_{12}(Mn,Fe)_3Si$; $Al_6(Mn,Fe)$	7x75	$Al_{12}Mg_2Cr$

2. Influence of Dispersoids on Fatigue Behavior

Dispersed second phase particles can play an important role in both crack initiation and crack propagation phases of fatigue failure, through contradictory roles. Second phase particles are typically hard intermetallics whose elastic and plastic deformation behavior is very different from that of the metallic aluminum matrix. Fine non-shearable particles, which cause dislocations to bypass them by looping help distribute the dislocations by promoting non-planar slip. This reduces dislocation pile-ups and thereby delay crack initiation [2]. On the other hand, shearable particles promote planar slip, which makes crack closure, crack branching, crack deflection and slip reversal possible during stress reversal, thereby decreasing the FCG rate [2]. Strain incompatibility between the particle and the matrix during mechanical loading could also initiate crack formation at the interface. Most commercial alloys contain a combination of both shearable precipitates and strong precipitates [6].

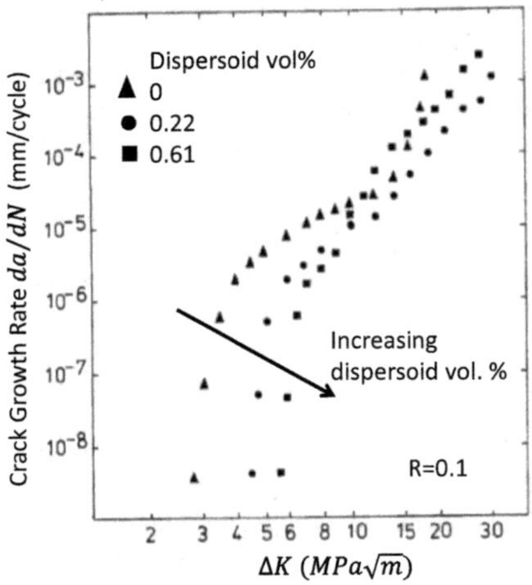

Figure 2. Effect of dispersoids content on ΔK_{th} [10]

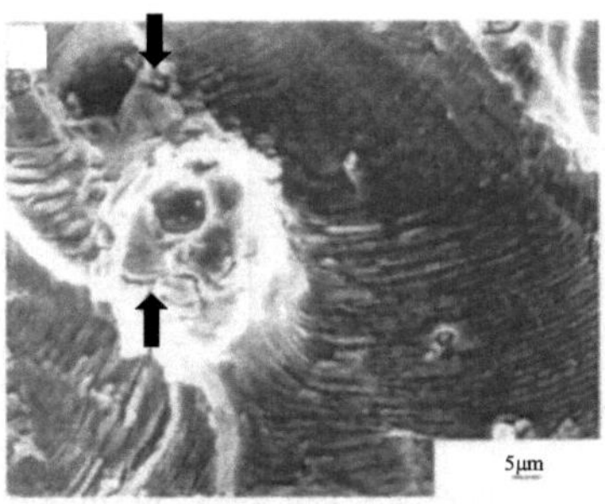

Figure 3: Cracked dispersoid indicated by black arrow [12]

2.1 Fatigue Crack Initiation

Cracks typically initiate at the surface of metallic components because stresses typically are the largest at the surface during, for example, bending or torsional loading. Typically, deformation is inhomogeneous and dislocation motion gets confined to persistent slip bands (PSB) in the presence of stress concentrations or precipitate free regions near grain boundaries. If the particles are shearable, dislocations can cut through the particles and reduce their size below the stability limit. This also leads to the formation of PSB and crack initiation [7,8].

It is generally accepted that strong second phase particles, whether they be precipitates during aging treatments or dispersoids, limit planar dislocation motion and distribute the deformation. This increases the strength, reduces the tendency to form

PSB, and decreases the propensity for crack initiation [2,9]. In a study on the influence of manganese addition to the ternary Al-Mg-Si alloys Edwards and Martin [10] show that Mn precipitates as $Al_{12}Mn_3Si$ dispersoids. The dispersoids were roughly equiaxed, about 0.1 μm in size. As the volume fraction of the dispersoids was increased, the deformation became more homogeneous. At the same time the threshold stress intensity range ΔK_{th} for initiation of fatigue cracks increased (Figure 2). In the absence of dispersoids, the grain size tends to be large, and cracks have been observed to nucleate at persistent slip bands (PSB) at grain boundaries [11].

However, strain incompatibility between the particle and the matrix can lead to interfacial de-bonding and crack formation. If the particle is brittle, fracture during prior manufacturing operations can lead to pre-cracking. An example of such pre-cracking is shown in Figure 3 from the work by Singh and Goel [12].

2.2 Fatigue Crack Propagation

The propagation of fatigue cracks in metals occurs by plastic deformation at the tip that results in progressive blunting and re-sharpening. This process can be thought of as an intrinsic damage process, i.e, the result of composition and microstructural characteristics of the alloy [2,13]. On the other hand, phenomena such as crack branching and crack deflection that result due to second phase particles are considered extrinsic mechanisms [6]. In their work on γ-Fe-Ni-Al alloys, Hornbogen and Zum-Gahr [14] conclude that planar slip retards FCG because slip reversal is possible when dislocation motion is confined to a single plane. Large grained metals therefore have slower crack propagation rates and improved fatigue properties. For aluminum alloys, this would mean strengthening mechanisms, such as the presence of hard second phase particles or grain boundary dispersoids that limit grain sizes promote a more rapid crack growth rate and could be detrimental to fatigue properties.

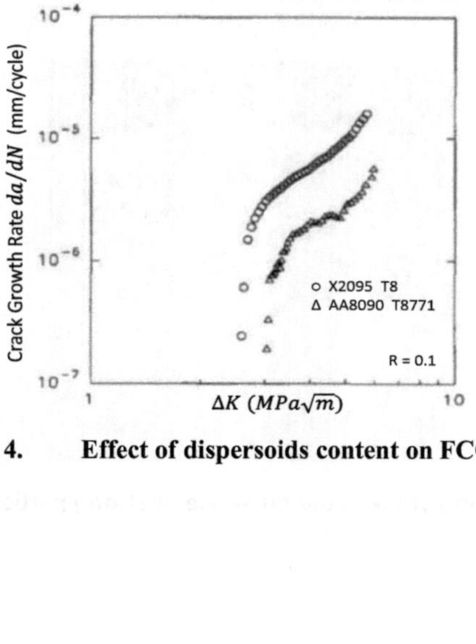

Figure 4. Effect of dispersoids content on FCG rate [5]

In Figure 4 Blankenship et al. [6] compare the behavior of two aluminum-lithium alloys, AA 8090 in the T8771 condition, and X2095 in the T6 condition. Both alloys contain hard particles – the T_1 (Al_2CuLi) phase in X2095 and S′ (Al_2CuMg) phase in AA8090. AA8090 is also strengthened by δ′ (Al_3Li) particles which can easily be sheared. At low values of the fatigue stress ratio ($R = \sigma_{min}/\sigma_{max} = 0.1$), the AA 8090-T8771 exhibits enhance crack growth resistance which was attributed to the coarse planar deformation which results in intrinsic microstructural damage mechanisms, such as crack branching, deflection, and void formation ahead of the crack tip [15] reduce crack growth rate. Chen et al. [16] show that in X2095, the homogeneously distributed T_B (Al_7Cu_4Li) particles caused fatigue cracks to deflect and meander resulting in a strong resistance to fatigue crack propagation in this alloy (Figure 5). By contrast, Albrecht et al. [17] found that Cr bearing dispersoid particles normally found in commercial AA 7075 fracture ahead of the crack tip and accelerate crack propagation. Telesman and Antolovich [18] investigated the effect of constant amplitude fatigue loading as well as variable amplitude fatigue or spectrum loading two alloys, relatively large grained AA 7050 produced by ingot metallurgy route and fine grained AA 7091 produced by powder metallurgy route. They show that at low stress intensities, the 7050 alloy showed better crack growth resistance than P/M 7091 alloy for both constant amplitude and spectrum testing. They suggest that large grain size is the most important microstructural parameter that accounts for low crack growth rates.

Warner et al. examined the effect of both grain size and dispersoid content on FCG rate in AA 2024 type alloys, at both low and high R values [19]. The general conclusions were that the FCG rate increases with increasing dispersoid density at a low R value of 0.05. The FCG rates tend to be smallest in low dispersoid content coarse grained material, and largest in high dispersoid content un-recrystallized material (Figure 6). These differences are, however, negligible at a high R value of 0.5 (Figure 7). Edwards and Martin [10] observed (see Figure 2) that for R=0.1, the FCG rate is smallest during Stage II for the dispersoid free alloy below a critical ΔK, but the alloy goes to Stage III at a lower ΔK, than those containing dispersoids.

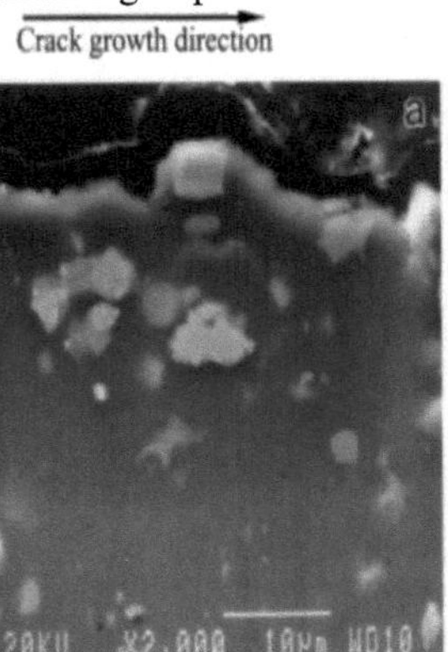

Figure 5. Meandering crack in the presence of strong particles [16]

In general, one can infer that grain size control through the introduction of dispersoids to prevent grain growth, is not desired from the point of view of fatigue crack growth rate.

2.3 Fracture Surface

The appearance of the crack surface depends upon the type of crack growth behavior and on ΔK. Near threshold, or Stage I of fatigue behavior, the microstructure failure mode is dominated by single slip, and the surface appears faceted because there is considerable meandering of the crack tip, shown schematically in Figure 8(a) [20]. During the Paris regime, or Stage II of fatigue behavior, multiple slip occurs and the fracture surface has a striated appearance (Figure 8(b)). During Stage III, K_{max} is large enough that static failure mechanisms take over and the fracture surface has a dimpled or faceted cleavage appearance.

Edwards and Martin [10] observed that for R=0.1, at a given ΔK, as the volume fraction of dispersoids increases, the degree of intergranular failure decreases (see Figure

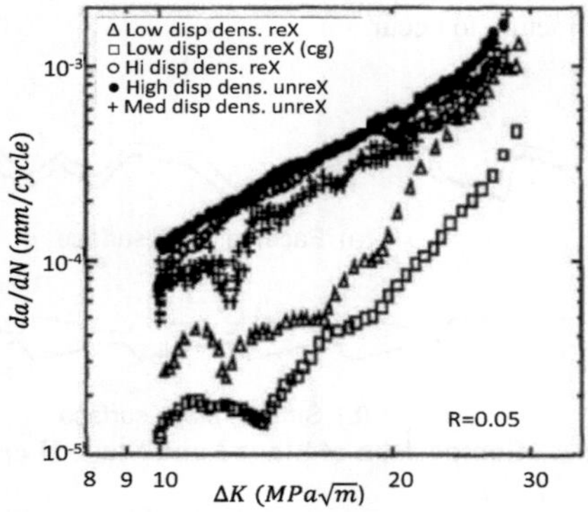

Figure 6. Effect of dispersoids content and grain size on FCG at R=0.05 [19]

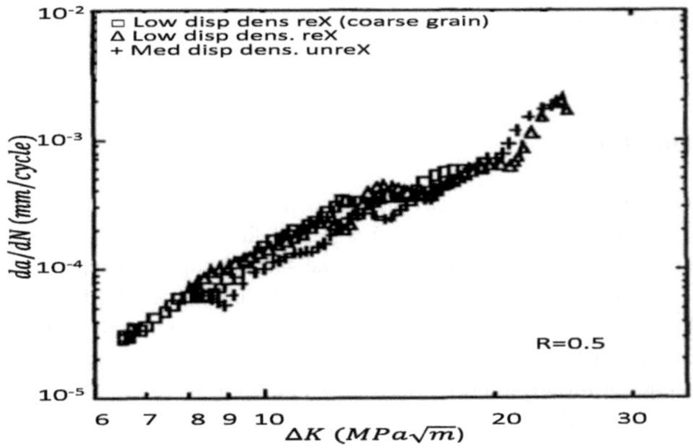

Figure 7. **Effect of dispersoids content and grain size on FCG at R=0.5**

9). This transition from intergranular to transgranular fracture has been attributed to the homogenization of slip by the dispersoids. Prince and Martin [21] rationalize this observation by proposing that there is a local minimum strain required for intergranular fracture. The homogenization of slip by the dispersoids leads to a larger macroscopic strain for intergranular fracture to occur.

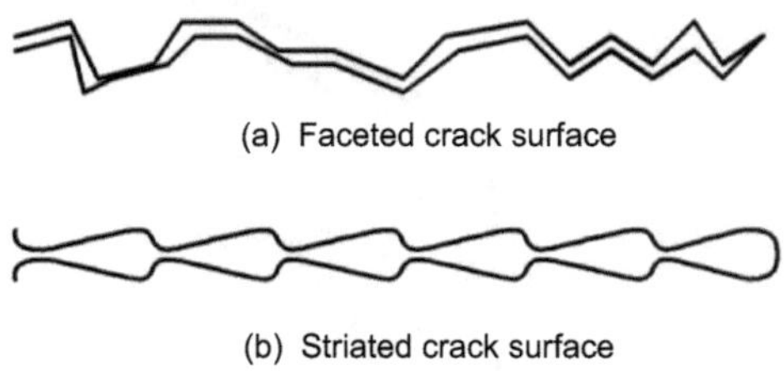

Figure 8. Comparison of Stage I and Stage II crack paths [20]

3. Summary

Commercial aluminum alloys typically contain second phase particles which are called precipitates, dispersoids, or inclusions, depending on their size and origin. Inclusions are typically oxide impurities and their presence can be controlled through process control. Precipitates and dispersoids, on the other hand, are intentionally introduced into the alloy to control microstructure and mechanical properties. Precipitates in age hardening (heat treatable) aluminum alloys pose barriers to dislocation motion, while dispersoids in both heat treatable and non-heat treatable alloys control grain size. Both these microstructural effects of second phase particles homogenize

(a) 0 vol% dispersoids

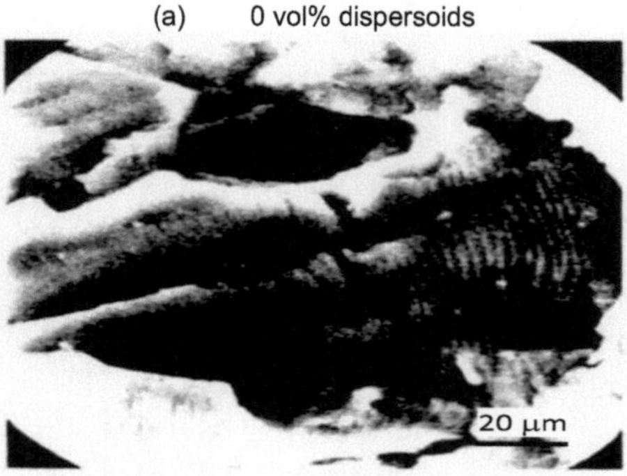

(b) 0.61 vol.% dispersoids

Figure 9. Increase in dispersoid volume fraction changes failure mode from inter-granular to trans-granular [10]

deformation and are beneficial from the point of view of the strength of the alloys. However, the effect of second phases, especially dispersoids, on fatigue behavior is more complex, depends upon a number of factors, including volume fraction, shape, size and distribution of the particles, and the type of fatigue loading being applied.

4. Concluding Highlights:

1. Dispersoids, like precipitates, can help in homogenizing deformation, and thus decrease the propensity for fatigue crack initiation. However, if dispersoids fracture during the processing steps prior to fatigue loading, they can become fatigue crack initiation sites.
2. Smaller grain sizes and reduced slip plane lengths due to the presence of dispersoids may decrease slip reversals during stress cycling. This can lead to increased FCG rates during Stage II fatigue. This effect may not be observed at high R values.
3. At low dispersoid volume fractions, intergranular fracture is observed. The fracture mode changes to transgranular fracture with increasing volume fractions of dispersoids.

References

[1] Fatigue of Materials, S. Suresh, Safari books online; Cambridge, U.K. ; New York : Cambridge University Press, 2004, c1998 01/01/2004

[2] C.P. Blankenship. Jr., E.A. Starke, Jr., and E. Hornbogen, "Microstructure and Properties of Aluminum Alloys," Chapter 1 in Microstructure and Properties of Materials, J.C.M Li Editor, World Scientific 1996

[3] Fatigue Crack Initiation and Propagation in Al-Alloys, Total Material http://www.keytometals.com/page.aspx?ID=CheckArticle&site=ktn&NM=197, accessed June 10, 2014

[4] A. D. Rollett, Lecture notes

[5] M. Tiryakioglu and J.T. Staley, "Physical Metallurgy and Effect of Alloying Additions in Aluminum Alloys," Chapter 3 in Handbook of Aluminum, Vol 1: Physical Metallurgy and Processes, G.E. Totten and D. Scott McKenzie, MarcelDekker, 2003.

[6] C. P. Blankenship, Jr., E. Hornbogen, and E. A. Starke, Jr, "Predicting slip behavior in alloys containing shearable and strong particles," **Materials Science and Engineering**, A pp 33-41, 1993.

[7] I.J. Polmear, "Solute Partitioning to Enhance Mechanical Properties of Aged Aluminum Alloys," Chapter 13 in Fundamentals of Aluminum Metallurgy, Production, Processing and Applications, edited by R Lumley, Woodhead Publishing, 2011.

[8] I.J. Polmear and R.N. Lumley, "Solute Partitioning to Enhance Mechanical Properties of Aged Aluminum Alloys," **Materials Science Forum,** Vols. 618-619, pp 487-491, 2009

[9] K.C. Kim and S.W. Nam, "Effects of Mn-dispersoids on the Fatigue Mechanism in an Al-Zn-Mg Alloy," **Materials Science and Engineering** A244, pp 257-262, 1998

[10] L. Edwards and J.W. Martin, "Effect of dispersoids on fatigue crack propagation in aluminium alloys," **Metal Science** Vol. 17, pp 511-518, 1983

[11] www.keytometals.com/page.aspx?ID=CheckArticle&site=ktn& NM=197, accessed June 10, 2014

[12] S. Singh and D.B. Goel, "Influence of thermomechanical aging on fatigue behaviour of 2014 Al-alloy," Indian Academy of Sciences **Bull. Mater. Sci**., Vol. 28, pp. 91–96, 2005

[13] R.O. Ritchie, “Mechanisms of Fatigue Crack Propagation in Metals, Ceramics and Composites: Role of Crack Tip Shielding,” **Materials Science and Engineering**, A 103, pp 15-28 1988

[14] E. Hornbogen and K-H Zum-Gahr, “Microstructure and Fatigue Crack Growth in a □-Fe-Ni-Al Alloy, **Acta Metallurgica**, Vol 24, pp 581-592, 1976

[15] R.O. Ritchie, “Mechanisms of fatigue-crack propagation in ductile and brittle solids, “, **Intl. J Fracture**, Vol 100, pp 55-83, 1999.

[16] D.L. Chen, M.C. Chaturvedi, N. Goel, N.L. Richards, ‘Fatigue crack growth behavior of X2095 Al–Li alloy,” **International Journal of Fatigue,** Vol 21 pp 1079–1086, 1999

[17] J. Albrecht, J.W. Martin, G. Lutjering and J.W. Martin, in 4th Int. Conf. on 'Strength of metals and alloys', LaboratoiredePhysiqueduSolide), Vol. 2, p463, 1976

[18] J. Telesman and S. D. Antolovich, “A Study of Spectrum Fatigue Crack Propagation in Two Aluminum Alloys II - Influence of Microstructures,” NASA Technical report, TM-86930 1985

[19] T. Warner, R. Dif, J.C. Ehstrom and P. Lassince, “Improved Durability Aluminum Alloys for Airframe Structures,” in the Proceedings of the RTO AVT Workshop on “New Metallic Materials for Structure of Aging Aircraft,” Corfu, Greece, 19-20 April 1999

[20] J.A. Newman, “The Effects fo Load Ratio on Threshold Fatigue Crack Growth of Aluminum Alloys,” PhD Dissertation, Virginia Polytechnic Institute and State University, 2000

[21] K.C. Prince and J.W. Martin, “The Effects of Dispersoids upon the Micromechanisms of Crack Propagation in Al-Mg-Si Alloys,” **Acta Metallurgica**, Vol 27, pp 1401-1408, 1979

Fatigue of Materials III: Advances and Emergences in Understanding
Edited by: T.S. Srivatsan, M. Ashraf Imam, and R. Srinivasan
TMS (The Minerals, Metals & Materials Society), 2014

A STUDY on FATIGUE CRACK CLOSURE BEHAVIOR in ALUMINUM ALLOY 7075 T6

Kenichi Masuda[1], Sotomi Ishihara[2], Yuki Tochikawa[1], Arthur J. McEvily[3], Masaki Okane[2], Seiichi Nishino[4]

[1] **University of Toyama**
3190 Gofuku Toyama 930-8555 **Japan**

[2] **National Institute of Technology, Toyama College**
13 Hongo Toyama 939-8630 **Japan**

[3]Department of Materials Science and Engineering
University of Connecticut
Storrs, CT, 06269, **USA**

[4] **National Institute of Technology, Anan College**
265 Aoki Minobayashi
Tokushima 774-0017 **Japan**

Abstract

In this study, fatigue tests were conducted on aluminum alloy 7075-T6 to study fatigue crack growth and crack closure behavior of the material. The experiments included the determination of the crack-opening levels K_{op} as a function of ΔK, the rate of fatigue crack growth da/dN as a function of ΔK_{eff}, where ΔK_{eff} (= K_{max}-K_{op}) is the effective stress intensity factor range. In addition, a change in the value of K_{op} due to the specimen-surface removal and a crack front shape were also investigated. It was concluded that the aluminum alloy 7075-T6 exhibits plasticity-induced fatigue crack closure. The crack growth rates and crack opening behavior were in good agreement with similar results in the literature.

Keywords: Fatigue crack propagation, Crack closure, Effective stress intensity factor

1. Introduction

In the 1970s, Elber [1, 2] introduced the important topic of fatigue crack closure into the literature and showed by means of experiments with the aluminum alloy 2024-T3 (yield strength [YS] = 350 MPa) that at a given R value, the crack-opening level increased with ΔK. Similar results have been obtained with other low- and medium-strength aluminum alloys. This type of crack closure is referred to as plasticity-induced fatigue crack closure (PIFCC), and Elber proposed that PIFCC was caused by the residual stretch of material present in the wake of the crack tip. Soon after Elber's initial studies, other forms of crack closure were found, including a type of material-related crack closure now known as roughness-induced fatigue crack closure (RIFCC). In this type of crack closure, the crack-opening level in the macro crack range is constant and independent of the ΔK level, in contrast to PIFCC. Schivje in his review article on crack closure [3] pointed out that the crack-opening level at the surface of a plane specimen, the plane-stress region, is higher than in the interior plane-strain region.

Newman et al. [4] found that the aluminum alloy 7075-T6 showed PIFCC behavior. Figure 1(a) shows the relations of ΔK, ΔK_{eff} and da/dN for Newman's results [4]. Based on Fig. 1(a), the K_{op} level can be derived as shown in Fig. 1(b). As can be seen from the figure, the crack-opening level increases with ΔK. This experimental result indicates that PIFCC controls the crack closure behavior in this material.

2. Material, Specimen and Experimental Procedures

The material tested was the aluminum alloy 7075-T6. Its chemical compositions and mechanical properties are listed in Tables 1 and 2, respectively.

ASTM-type compact specimens [5] were used for the crack growth studies. Figure 2 shows the geometric shape of the specimen. The specimens with W equal to 57.2 mm and thicknesses equal to 6 mm were used in the tests. The fatigue crack growth (FCG) tests were carried out in a servo controlled hydraulic test system under load control condition in laboratory air at room temperature under stress ratio of R = 0.1. The test frequency was 15 Hz. Crack lengths were measured using the replication technique. The elastic compliance method was used for a determination of the crack-opening level. In the current study, the elastic compliance method using the subtracted strain was used to improve the sensitivity and precision of the determination of the crack-opening load.

In the surface-removal experiments, a fatigue crack was propagated under a constant stress-intensity factor range ΔK. After reaching the steady-state level of fatigue crack closure, the specimen was taken from the fatigue testing machine, and then specimen surfaces were removed by 0.5 mm each using an electric discharge machine. After surface removal, the fatigue crack propagation test was resumed and the K_{op} level was determined anew.

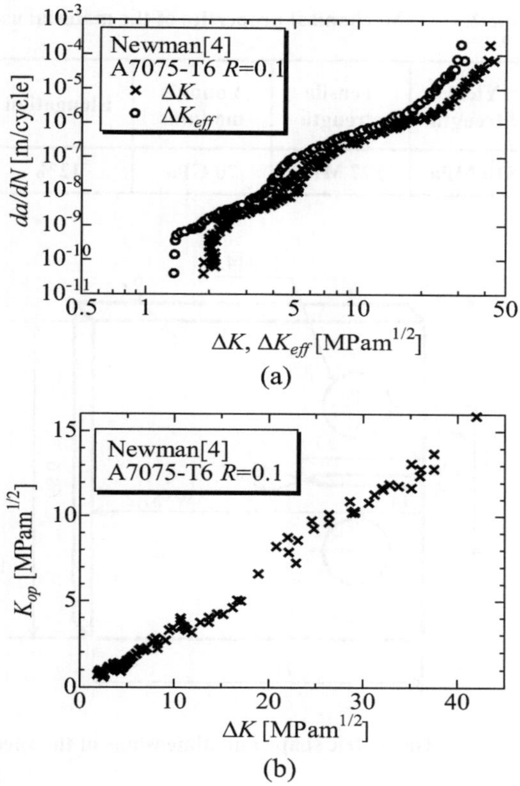

Figure 1. Newman's results [4] for:
(a) *da/dN* vs. ΔK, ΔK_{eff} relations; (b) K_{op} vs. ΔK relation.

Table 1. Chemical compositions of the material used (weight percent)

Si	Fe	Cu	Mn	Mg	Cr	Zn	Ti	Al
0.4	0.5	1.6	0.3	2.5	0.24	5.5	0.2	Bal.

Table 2. Mechanical properties of the material used

Yield Strength	Tensile strength	Young's modulus	Elongation
510 MPa	**577 MPa**	**70 GPa**	**12%**

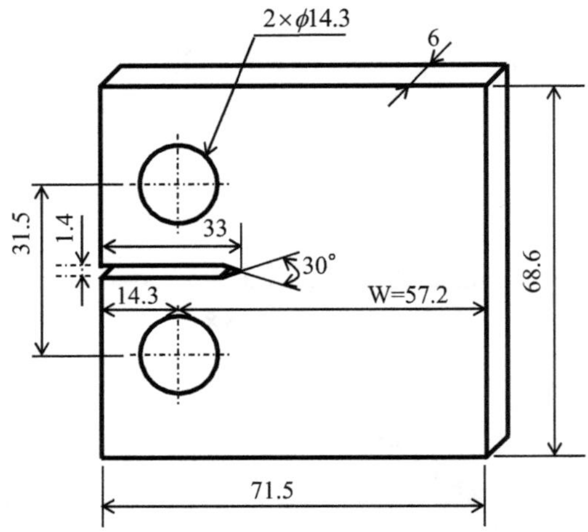

Figure 2. **Geometric shape and dimensions of the specimen**

3. Experimental Results and Discussion

3.1 Fatigue Crack Propagation and Crack Closure

It is of interest to know how the results of one investigation compare with those from another. Figure 3(a) shows the comparison between present results and Newman's results for vK, ΔK_{eff} and da/dN relations. Figure 3(b) shows the relation of K_{op} and ΔK. As you can see, the crack-opening level increases with ΔK in accord with Newman's findings, which confirms that PIFCC is dominant in these tests. Also, as seen from these figures, the present results are in good agreement with Newman's results.

3.2 Effect of Surface Removal

In this experiment a fatigue crack was grown in a CT specimen at a ΔK level of 8.0 $MPam^{1/2}$. The specimen was then removed from the testing machine and 0.5 mm was removed from each specimen surface by electro discharge machining. Then both surfaces were electro polishing. After the surface removal, the crack was advanced further by cyclic loading and crack closure levels were determined. Figure 4 shows how the crack-opening level, K_{op}, varied with crack length, both before and after surface removal.

This result showed a marked effect of surface removal on the crack-opening level of A7075-T6, with the crack closure level at the surface being 1.0 $MPam^{1/2}$ higher than that in the interior. Because A7075-T6 exhibits PIFCC as shown in Fig. 3, the plane-stress plastic zone at the specimen surface is an important factor in PIFCC [6]. Further if the thickness of the specimen increases were to be increased in steps beyond that of the present 6 mm, it is expected that K_{op} would eventually drop to 2 $MPam^{1/2}$. Therefore there would be an effect of thickness on the fatigue crack growth rate, i.e., the rate would increase up to the thickness where RIFCC would become dominate, after which there would no longer be a thickness effect.

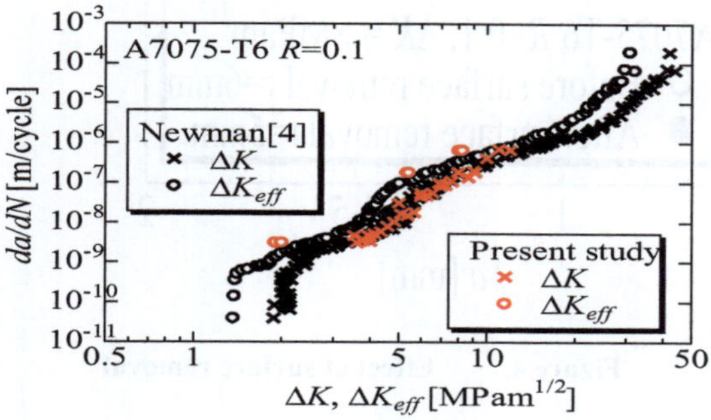

(a)

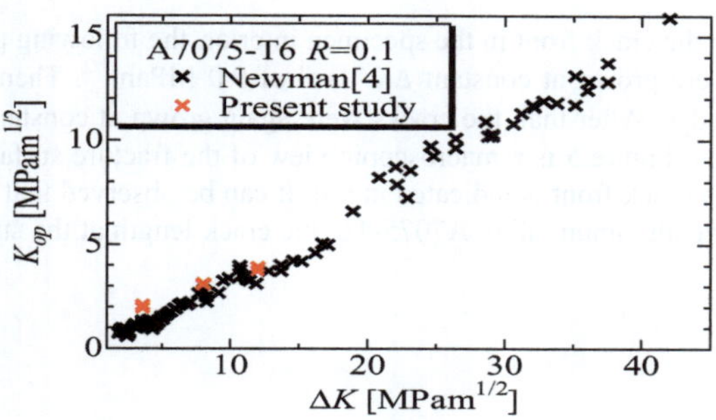

(b)

Figure 3. **Comparison between present results and Newman's results for:**
(a) **ΔK, ΔK_{eff} and da/dN relations;**
(b) **K_{op} and ΔK relation.**

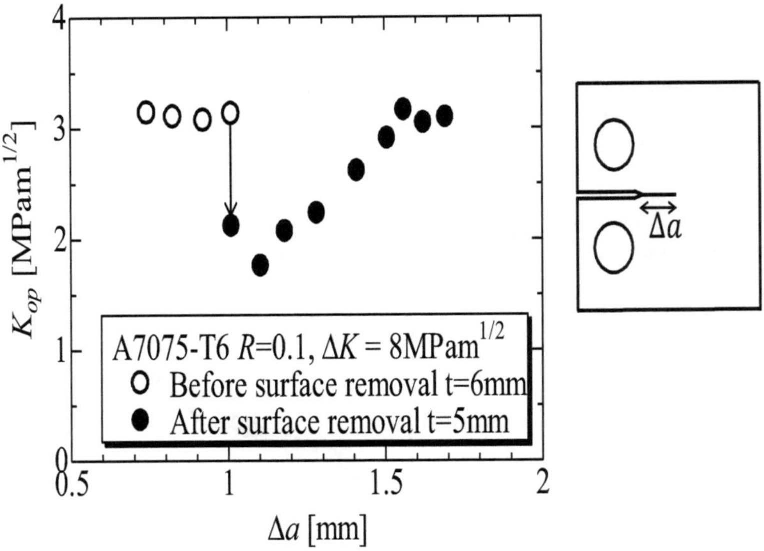

Figure 4. **Effect of surface removal**

3.3 Shape of the Crack Front

For identifying the crack front in the specimen interior, the following procedure was used. At first, the cracks were grown at constant ΔK level of 9.0 $MPam^{1/2}$. Then, the specimen was dipped in penetrating dye. After that, the cracks were again grown at constant ΔK level until the failure of the specimen. Figure 5 is a macroscopic view of the fracture surface of the specimen, where the shape of the crack front is indicated in red. It can be observed that because of a higher degree of PIFCC in the aluminum alloy A7075-T6, the crack length at the surface is less than in the interior.

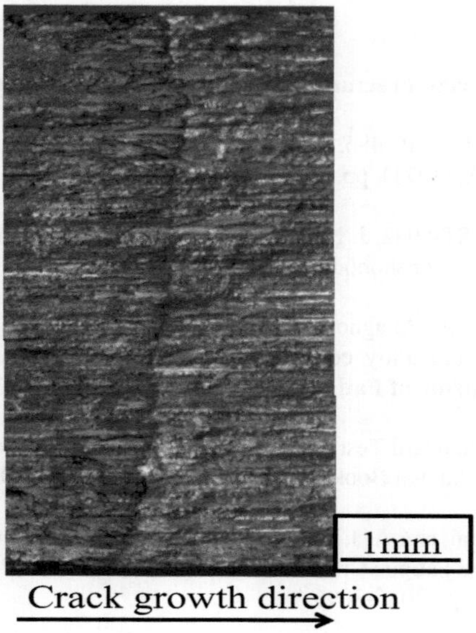

Figure 5. Macroscopic view of the shape of the crack front. Aluminum Alloy 7075-T6

4. Conclusions

The aluminum alloy 7075-T6 in a CT specimen of 6mm thickness exhibits plasticity-induced fatigue crack closure, PIFCC. The evidence is as follows:

1. The crack-opening level K_{op} increases linearly with the stress intensity factor range, ΔK.

2. An effect of surface removal is observed on the crack-opening level.

3. The crack length at the specimen surface is less than in the interior.

References

[1] W. Elber, **Engineering Fracture. Mechanics**, (1970), vol. 2, pp. 37-45.

[2] W. Elber, ASTM STP 486, American Society for Testing and Materials, West Conshohocken, PA, (1971), pp. 230-42.

[3] J. Schivje, ASTM STP 982, J. Newman and W. Elber, eds., American Society for Testing and Materials, West Conshohocken, PA, (1988), pp. 5-34.

[4] J.C. Newman Jr., E.L. Anagnostou and D. Rusk, Fatigue and crack-growth analyses on 7075-T651 aluminum alloy coupons under constant- and variable-amplitude loading, **International. Journal of Fatigue,** (2014), Vol. 62, pp.133-143.

[5] ASTM E399-90, Standard Test Method for Plane-Strain Fracture Toughness of Metallic Materials, (1991), Annual Book of ASTM Standards, pp. 500-02.

[6] S. Ishihara, Y. Sugai, and A. J. McEvily, On the Distinction Between Plasticity- and Roughness-Induced Fatigue Crack Closure, **Metallurgical and Materials Transactions A**, (2012), Vol. 43, pp.3086-3096.

FATIGUE of MATERIALS III
ADVANCES and EMERGENCES in UNDERSTANDING

Advanced Materials

Fatigue of Materials III: Advances and Emergences in Understanding
Edited by: T.S. Srivatsan, M. Ashraf Imam, and R. Srinivasan
TMS (The Minerals, Metals & Materials Society), 2014

FATIGUE BEHAVIOR OF ULTRA-FINE GRAINED Ti-6Al-4V ALLOY

R. Ebara[1], M. Endo[1,2], H.J. Kim[1,2], J. Nakahigashi[3]

[1] *Institute of Materials Science and Technology*
Fukuoka University
8-19-1, Nanakuma, Jonan-ku,
Fukuoka-city, 814-0180, **Japan**
E-Mail:ebara@fukuoka-u.ac.jp

[2] *Department of Mechanical Engineering*
Fukuoka University
8-19-1, Nanakuma, Jonan-ku,
Fukuoka-city, 814-0180, **Japan**

[3] *Department of Mechanical System Engineering*
Fukuyama University
1-3, Gakuen,
Fukuyama-city, 729-0292, **Japan**

Abstract

This paper reviews fatigue strength, fatigue crack initiation and fatigue crack propagation behavior of protium treated ultra-fine grained Ti-6Al-4V alloy in air and in 3% NaCl aqueous solution. First the protium treatment procedure for Ti-6Al-4V alloy and characteristics of ultra- fine grained Ti-6Al-4V alloy such as microstructure and mechanical properties is briefly summarized. Then the fatigue and corrosion fatigue strength of ultra-fine grained Ti-6Al-4V alloy are described mainly on the basis of the authors' recent experimental findings. The emphasis is placed upon fatigue crack initiation and propagation behavior of ultra-fine grained Ti-6Al-4V alloy in air and in 3% NaCl aqueous solution. Finally a replica method expected to reveal crack initiation and early crack propagation mechanism in ultra-fine grained Ti-6Al-4V alloy is described by demonstrating a couple of recent findings on conventional Ti-6Al-4V alloy.

Keywords: Protium treatment, Ultra-fine grained Ti-6Al-4V alloy, Fatigue ,Crack initiation and propagation.

1. Introduction

Recent progress of grain refinement technique significantly increased fatigue strength of structural materials such as various kinds of steel [1-4], Ti-6Al-4V alloy [5-10] and Aluminium alloy [11, 12]. Ti-6Al-4V alloy with excellent properties such as specific strength, heat resistance and corrosion resistance has been widely used in various industries. The grain refinement technique has brought the possibility of extensive use of Ti-6Al-4V alloy with ultra-fine grain size. One of the authors (Nakahigashi) has been concerned on protium treatment of Ti-6Al-4V alloy and has successfully found a superplasticity in protium treated Ti-6Al-4V alloy for denture bases [13-15]. A beneficial effect on fatigue properties was also reported on Ti-6Al-4V ELI alloy for medical use [9]. The higher fatigue properties of ultra-fine grained Ti-6Al-4V alloy are indispensable for extensive use of this alloy in various applications. This paper reviews fatigue strength, i.e., fatigue crack initiation and propagation behavior, of protium treated ultra-fine grained Ti-6Al-4V alloy in air and in 3% NaCl aqueous solution mainly on the basis of the authors' recent experimental findings [7, 10].

2. Ultra-fine grained Ti-6Al-4V alloy

Ultra-fine grained Ti-6Al-4V alloy can be obtained by protium treatment developed by one of the authors (Nakahigashi) [13-15]. Protium treatment procedure is schematically shown in Figure 1.Plates with 25 mm wide, 75 mm long and 30 mm thick were cut from rolled Ti-6Al-4V alloy with conventional chemical compositions. The plates were exposed under hydrogen environment at 1073K for 1.8 ks and were cooled in air. The amount of protium absorbed in the plate was 0.5 mass%. Then the plates were heated up to 1223K above β transformation temperature in air and were cooled in water. The plates were reheated up to 1023K and were rolled with reduction rate of 80 percent in air, then cooled in water. Finally, in order to eliminate absorbed hydrogen, the plates were kept for 10.8ks at 873K under vacuum and were cooled in furnace. Throughout this process ultra-fine grained Ti-6Al-4V alloy with grain size of 0.5 μm can be obtained. Figure 2 a) shows microstructure obtained by optical microscopy of ultra-fine grained Ti-6Al-4V alloy prepared for after mentioned axial fatigue test. Grain size is 0.5 μm and is less than one tenth of that of conventional Ti-6Al-4V alloy whose grain size is 5.1 μm (Figure 2 b)).

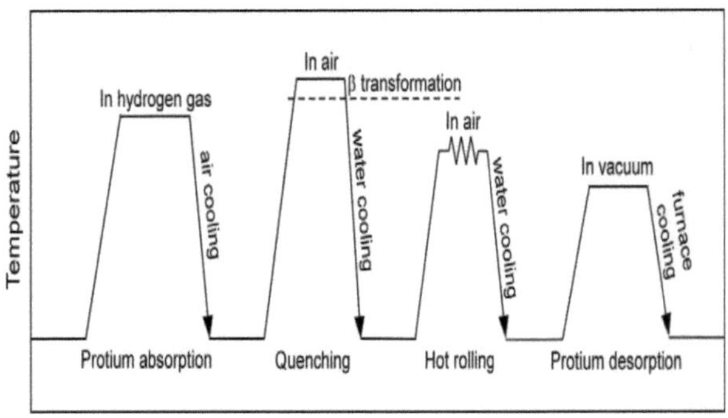

Figure 1. Schema of protium treatment process [7].

If the materials were treated by a process in which 0.5 mass% of protium is absorbed and desorbed at 973K, a fine grain size of about 1-3 μm were obtained. Subsequently the desorption temperature was decreased to 873K, the materials began to exhibit an ultra-fine and equiaxial grains as shown in Figure 3 [16]. TEM observations also revealed the grain-refined structure with grain size from 0.3 to 0.5 μm. According to SEM and TEM observation, ultra-fine microstructures formed in protium treated process was consisted of three stages of (1) protium precipitation in grain, (2) grain formation in the matrix through desorption and (3) formation of ultra-fine grain. Furthermore the protium treatment increased β-phase percentage in α-phase matrix. The proportion of the β-phase matrix was about 11% in a fine grained and 14% in an ultra-fine grained material, respectively [16]. The most beneficial feature in ultra-fine grained Ti- 6Al-4V alloy is its higher proof stress and ultimate tensile stress than those of conventional Ti- 6Al-4V alloy. Figure 4 shows the stress-strain curves of Ti-6Al-4V alloys with different grain size. The ultimate tensile strengths of ultra-fine grained Ti-6Al-4V alloy, fine grained Ti-6Al-4V alloy and coarse grained Ti-6Al-4V alloy were 1266,

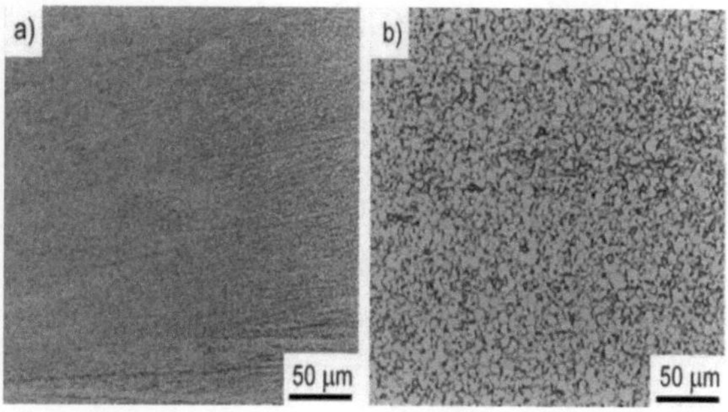

a) Ultra-fine grain b) Conventional grain

Figure 2. Microstructure of ultra-fine grained Ti-6Al-4V alloy [7].

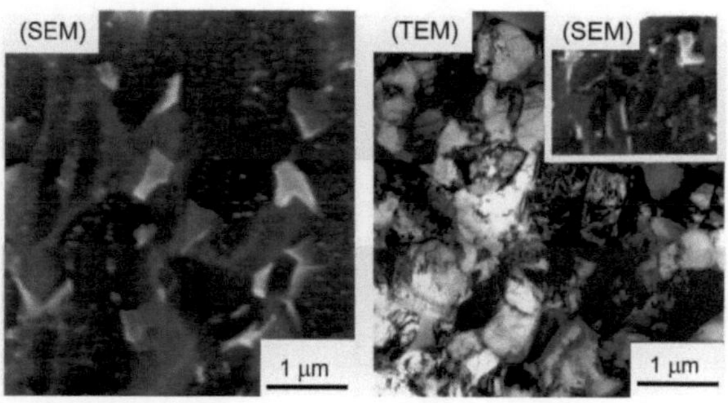

a) Fine grain b) Ultra-fine grain

Figure 3. SEM and TEM micrographs of microstructures of protium treated Ti-6Al-4V alloy [16].

1127 and 1026 MPa, respectively. The smaller the grain size the higher the ultimate tensile strength is. The Hall-Petch relation was observed between 1.2 % proof stress and negative half-power of the grain size of Ti-6Al-4V alloys. In contrast the smaller the grain size, the smaller the strain was [7]. Dimple was predominantly observed on the tensile fracture surfaces of these Ti-6Al-4V alloys with different grain sizes. The dimple size was approximately equivalent to each grain size of the Ti-6Al-4V alloys with different grain sizes as shown in Figure 5 [17].

One of the prominent mechanical properties of ultra-fine grained Ti-6Al-4V alloy is superplasticity. As shown in Figure 6, superelongation over than 6000% is observed in ultra-fine grained Ti-6Al-4V alloy. Application of ultra-fine grained Ti-6Al-4V alloy is expected in various fields including dental material.

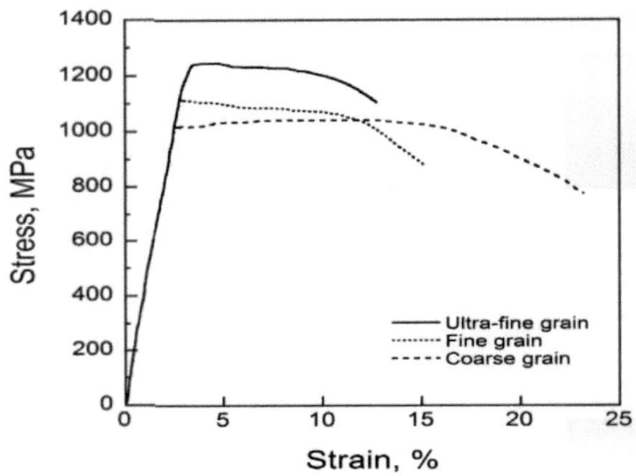

Figure 4. **Stress versus strain crves of Ti-6Al-4V alloy with grain sizes [7]**

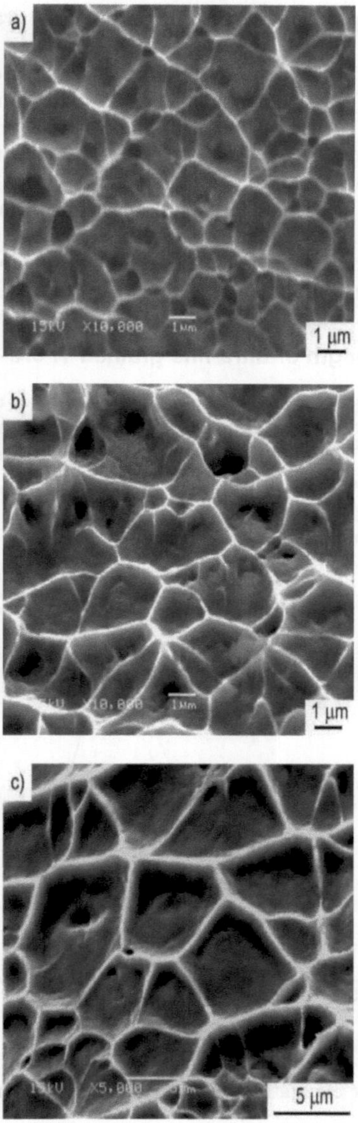

Figure 5. Dimples on tensile fracture surface [17]; a) Ultra-fine grain b) Fine grain (c) Coarse grain.

3 Fatigue behavior of ultra-fine grained Ti-6Al- 4V alloy

3.1 Rotating bending fatigue strength of ultra-fine grained Ti-6Al-4V alloy

Rotating bending fatigue tests were conducted by use of round bar specimen with 10 mm diameter [5]. Fatigue strength of ultra-fine grained Ti-6Al-4V alloy with a grain size of 1 μm at 10^7 cycles was 600 MPa and was 50% higher than that of Ti-6Al-4V alloy with a grain size of 16 μm. The similar rotating bending fatigue test results revealed that the Ti-6Al-4V alloy with a grain size of 1 μm was 50% higher than that of Ti-6Al-4V alloy with a grain size of 20 μm [6]. Thus it can be concluded that fatigue strength of the ultra-fine grained Ti-6Al-4V alloy can be improved by protium treatment. Granular fracture surface was predominant in fatigue crack propagation area of ultra-fine grained Ti-6Al-4V alloy. Striation was partially observed on crack propagation area. On the contrast, a striation was predominant on Ti-6Al-4V alloy with a grain size of 16 μm [6].

3.2 Axial fatigue strength of ultra-fine grained Ti-6Al-4V alloy

Axial fatigue tests were conducted by use of the above mentioned protium treated Ti-6Al-4V alloy with grain sizes of 0.5 and 1.4 μm and conventional Ti-6Al-4V alloy with a grain size of 5.1 μm [7]. The ultimate tensile strength of ultra-fine grained with 0.5 μm, fine grained with 1.4 μm and coarse grained with 5.1 μm were 1125, 1071 and 955 MPa, respectively (Figure 4). Load controlled fatigue tests were conducted by use of the plane plate specimen with 20 mm

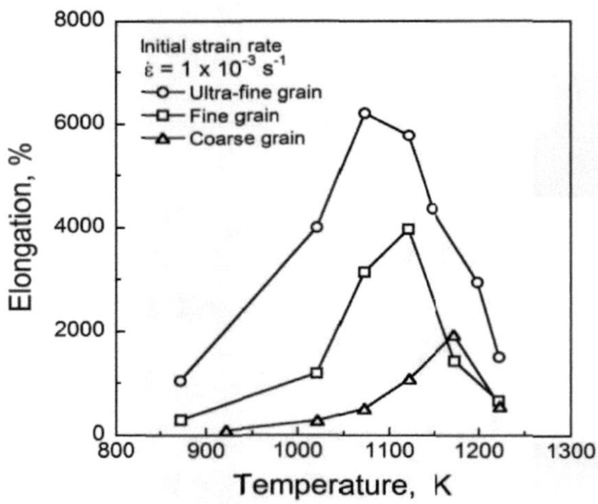

Figure 6. Superplastic elongation of Ti-6Al-4V alloys with different grain sizes [16].

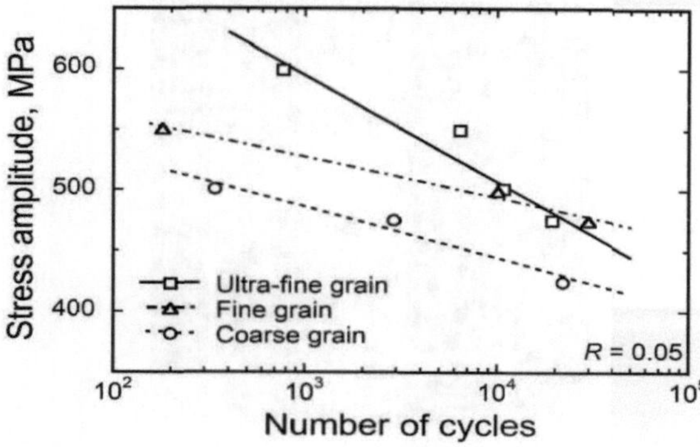

Figure 7. ***S-N* curves of Ti-6Al-4V alloy with different grain size in low cycle region [7].**

wide, 75 mm long and 4 mm thick at room temperature in air. The minimum width and length of the parallel section was 10 mm and 28 mm, respectively. Frequency was 20 Hz and *R* value was 0.05. Figure 7 shows *S-N* diagrams in low cycle fatigue region of Ti-6Al-4V alloys with different grain sizes. The smaller the grain size the higher the fatigue strength is. The grain size effect on fatigue strength is clearly observed at the number of cycles lower than 3×10^4 cycles. The higher the number of cycles the grain size effect on low cycle fatigue strength became smaller.

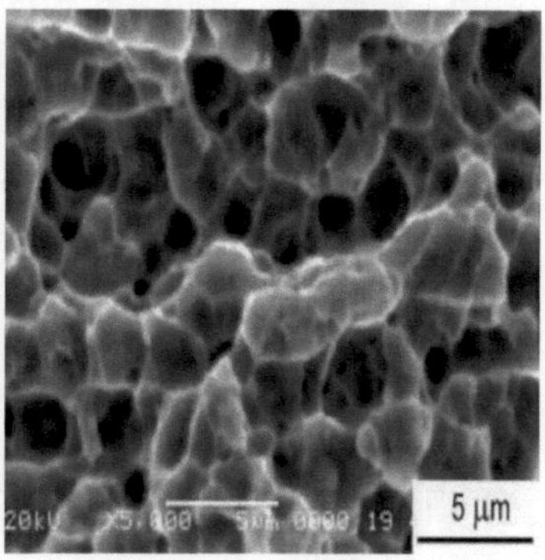

Figure 8. **Granular fracture surface observed in ultra-fine grained Ti-6Al-4V alloy [17].**

The fatigue life at a stress amplitude of 500 MPa for the ultra-fine grained Ti-6Al-4V alloy is thirty times longer than that of Ti-6Al-4V alloy with coarse grain. Fatigue crack initiated from specimen surface in all tested Ti-6Al-4V alloys with different grain size. Granular fracture surface was predominant in the crack propagation area. The role of ultra-fine grain in crack propagation stage is quite understandable in view of a typical granular fracture surface of the ultra-fine grained Ti-6Al-4V alloy as shown in Figure 8. Figure 9 shows low cycle fatigue fracture surfaces of Ti-6Al-4V alloy with different grain sizes.

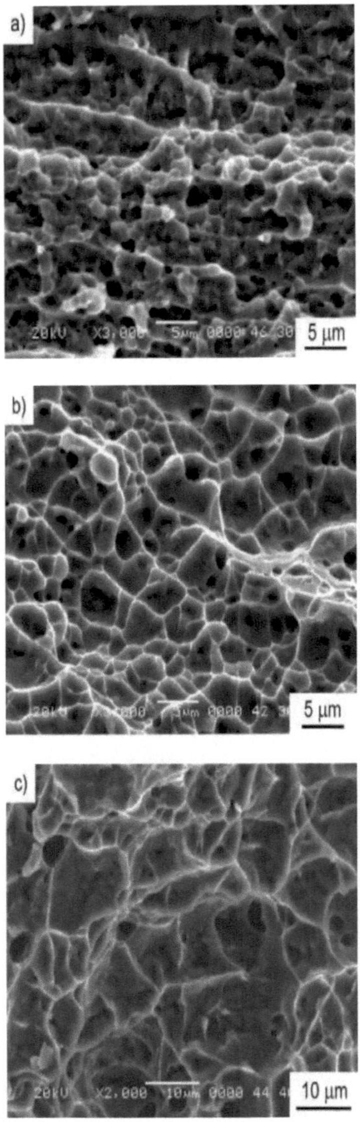

Figure 9. **Low cycle fatigue fracture surface of Ti-6Al-4V alloy with different grain size [17]. (a) Ultra-fine grain, (b) Fine grain, (c) Coarse grain.**

The size of the dimple observed on tensile fracture surface of each Ti-6Al-4V alloy is approximately equivalent to that of grain size as shown in Figure 9. It can be mentioned that the low cycle fatigue strength of Ti-6Al-4V alloy is strongly related to the ultimate tensile strength of Ti-6Al-4V alloy. The grain size effect on fatigue strength in low cycle fatigue region was also observed in high strength maraging steels with grain sizes of 20, 60 and 100 μm [18] and ferritic stainless steels with a grain size of 0.2 μm [4].

Figure 10 shows *S-N* diagrams of Ti-6Al-4V alloys with different grain sizes including that of low cycle fatigue region shown in Figure 7. It is clear that the fatigue strength of the ultra-fine grained Ti-6Al-4V alloy is higher than those of the fine grained and the coarse grained Ti-6Al- 4V alloys in high cycle fatigue region as well as in low cycle fatigue region. The fatigue strength of ultra-fine grained Ti-6Al-4V alloys at 10^7 cycles is 400 MPa and is about 1.2 and 1.7 times higher than those of fine and coarse grained Ti-6Al-4V alloys.Fatigue crack propagation test by use of the single round edge notched specimens with stress concentration factor of 2.5 showed that fatigue crack initiation considerably delayed in the ultra- fine grained Ti-6Al-4V alloy as compared with those of fine and coarse grained Ti-6Al-4V alloy. However, little difference was found on crack propagation rate of these Ti-6Al-4V alloys [17]. From this fact, it can be concluded that a high fatigue strength of ultra-fine grained Ti-6Al-4V alloy is due to the delay of crack initiation.

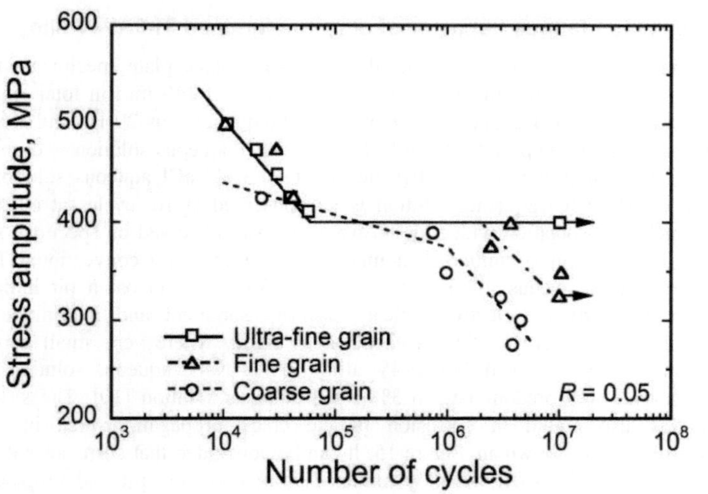

Figure 10. ***S-N*** **diagrams of Ti-6Al-4V alloys with different grain sizes [7].**

The proportion of the β-phase in ultra-fine grained Ti-6Al-4V alloy is higher than those in fine and coarse grained Ti-6Al-4V alloy. Therefore it can be summarized that the dispersed strain concentration to the β phase in ultra-fine grain possibly delay crack initiation and thereby increases the fatigue strength of ultra-fine grained Ti-6Al-4V alloy. Granular fracture surface was predominant in the high cycle fatigue region of ultra-fine grained Ti-6Al-4V alloy as shown in Figure 11 a). Aspect of granular fracture is more distinct in Figure 11 b). Striation like fracture surface and secondary cracks were observed very rarely as shown in Figure 11 c). Figure 12 shows representative high cycle fracture surfaces of Ti-6Al-4V alloys with different grain sizes. The similar granular fracture surface to that of ultra-fine grained Ti- 6Al-4V alloy was predominant in fine grained Ti-6Al-4V alloy (Figure 12 b)). On the contrary, striation was predominant in coarse Ti-6Al-4V alloy (Figure 12 c)). Fractographic features in single edge notched specimen for crack propagation test was basically the same as those in smooth specimen used for fatigue life test. The granular fracture surface was predominant in ultra-fine grained Ti-6Al-4V alloy. Striation was very rarely observed in granular fracture surface as shown in Figure 13.

The similar fractographic feature that striation partially observed in the restricted area of granular fracture was found on the fracture surface in ferritic stainless steels with ultra-fine grain of 0.2 μm in grain size [4] and in the pure Ti severely deformed by accumulative roll bonding process [19]. These ferritic stainless steels such as Fe-2.25Cr-1Mo- 1Zr, Fe-13Cr-8Ni-2Moo-1Zr, and Fe-25Cr-1Zr extruded after mechanical alloying had higher ultimate tensile strength and axial fatigue strength than those of conventional SAE8740 steel. The phenomenon of the partial appearance of striation or striation like pattern in granular fracture seems to be closely related to heavily deformed materials with high strength. The microscopic brittleness was also reported on VT9 titanium alloy with an ultimate tensile strength of 1350 MPa. The facets of brittle fracture surrounded by dimpled relief were observed [20].

3.3 Corrosion fatigue behavior of ultra-fine grained Ti-6Al-4V alloy

Axial fatigue tests were conducted by use of a plane plate specimen with 4 mm wide, 2.5 mm thick, 20 mm long of parallel section and 140 mm in total length in air and in aerated 3% NaCl aqueous solution at RT. Frequency was 20 Hz and R value was 0.05. In higher stress amplitude, the effect of 3% NaCl aqueous solution is observed. The lower the stress amplitude the smaller the effect of 3% NaCl aqueous solution is. The influence of 3% NaCl aqueous solution is not observed at 10^7 cycles at all. Corrosion fatigue crack in association with tiny corrosion pit was observed on specimen surface as shown in Figure 14. The similar phenomenon was observed for conventional Ti-6Al-4V alloy in 3% NaCl aqueous solution at 353K [21]. However, corrosion pit initiation was not detected clearly by electrochemical noise measurement and intermittent surface observation. Further investigation is expected to clarify where very small corrosion pit initiates in ultra-fine grained Ti-6Al-4V alloy in 3% NaCl aqueous solution. Granular fracture surface was predominant in 3% NaCl aqueous solution [10]. The striation like pattern was also found in corrosion fatigue crack propagation area in 3% NaCl aqueous solution as shown in Figure 15. It can be concluded that corrosion fatigue crack of ultra-fine grained Ti-6Al-4V alloy initiated from corrosion pit and propagated with granular fracture surface in 3% NaCl aqueous solution.

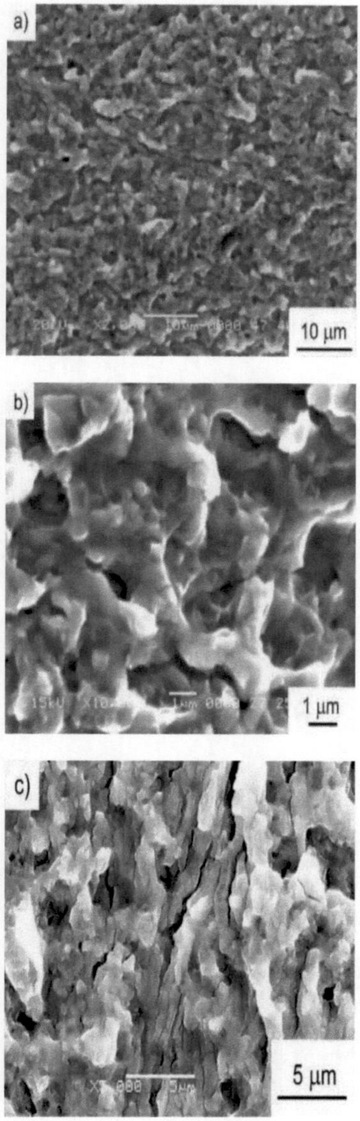

Figure 11. **Fatigue fracture surface morphology in ultra-fine grained Ti-6Al-4V alloy [7, 17].**

σa = 425MPa, N_f = 3.99 × 10^4 cycles.

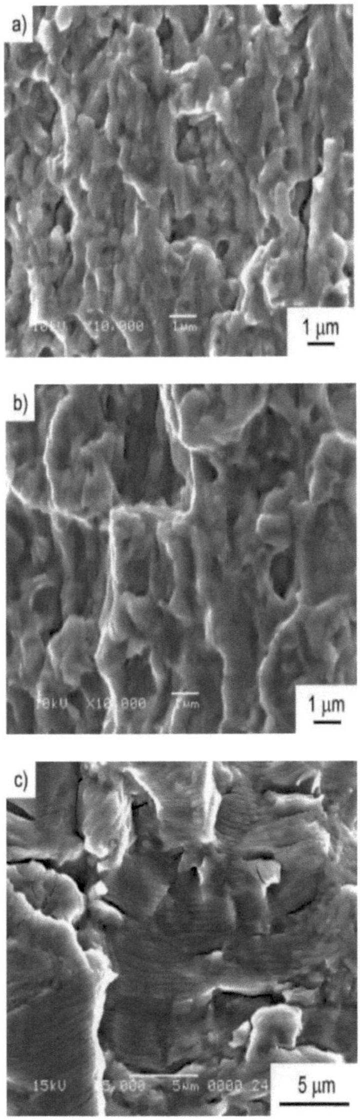

Figure 12. **Fatigue fracture surface morphology of Ti-6Al-4V alloy with different grain size [17]. (a) Ultra-fine grain, (b) Fine grain, (c) Coarse grain.**

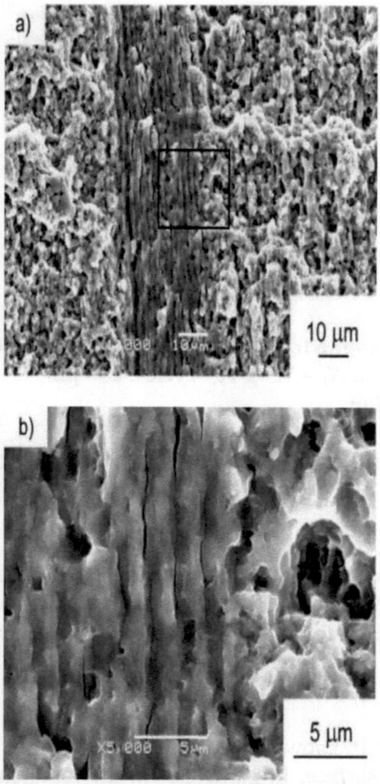

Figure 13 **Granular fracture surface and striation observed in crack propagation area of ultra-fine grained Ti-6Al-4V alloy [7].**
(a) Granular fracture surface and striation
(b) Magnified micrograph framed part in a).

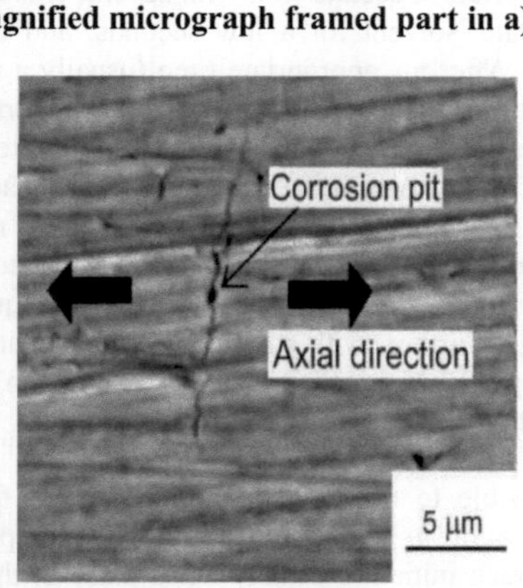

Figure 14. **Corrosion fatigue crack in association with corrosion pit [10]. 3% NaCl aqueous solution,**
σa = 475 MPa, Nf = 7.3x 10^3 cycles.

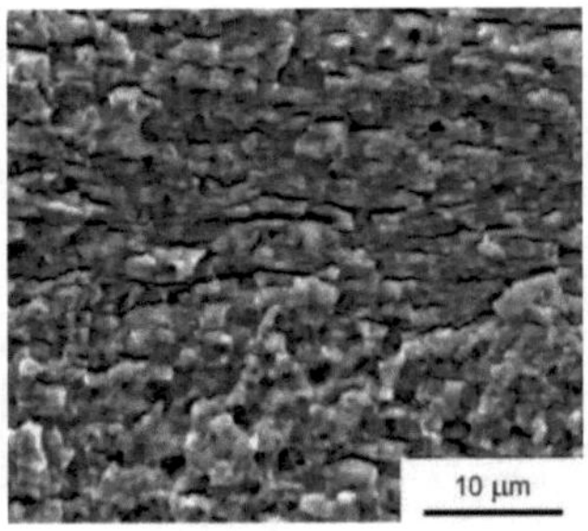

Figure 15. **Corrosion fatigue fracture surface [10]; 3% NaCl aqueous solution σ_a = 475 MPa, N_f = 7.3 x 10^3 cycles, 250 μm from initiation.**

3.4 Crack initiation and early propagation of small fatigue cracks in Ti-6Al-4V alloy

As aforementioned, study on the fatigue property of Ti-6Al-4V alloy with fine or ultra-fine grains has been progressed in recent years. However, the fatigue mechanism has yet to be sufficiently known. One of the best ways to investigate the fatigue mechanism is to observe successively the process of crack initiation and the early propagation behavior of a small fatigue crack determining dominantly fatigue lifetime and threshold.

The replica method is known as an effective means to investigate the microscopic morphology of material's surface. In particular, in the research field of metal fatigue, this method has often been used to make a successive observation of fatigue process on the surface of a specimen. The general procedure to take advantage of this method is described below. A small piece of cellulose acetate film with several tens of micrometers in thickness is immersed in methyl acetate solvent for a few seconds, and then the gelled film is put onto the specimen surface. After an appropriate time (usually a few minutes), the dried-out film is peeled off from the surface. This provides an inverted replication of surface topography. A series of replica films are taken by halting the testing machine at appointed numbers of load cycles. The whole fatigue process from crack initiation to specimen failure can successively be investigated by observing the replicas under an optical microscope. The transparent replica film just obtained is inadequate for observation, and thus gold deposition, for instance, by the ion sputtering technique is usually performed to obtain a clear surface image. Further, with additional proper treatments for replica film, the scanning electron microscopy (SEM) or the transmission electron microscopy (TEM) is also available to examine the fatigue process at higher resolution.

It is usually impossible to observe the initiation of a fatigue crack and its early propagation on the surface of a Ti-6Al-4V alloy smooth specimen in the high-cycle fatigue regime since the crack initiation usually takes place in the interior of the specimen, typically at a facet [22-24]. In the case of low-cycle fatigue, few fatigue cracks will appear on the smooth surface and usually a large number of replicas must be taken from a broad surface area to detect a dominant crack that leads to specimen failure, which results in a time-consuming task. Concerning Ti-6Al-4V alloy, the α-phase that plays a dominant role to fatigue crack initiation has a HCP structure with very few slip systems, so that only a very small number of grains are favorable for crack initiation on the specimen surface. It is noted that this is quite different from the case of steels or other alloys in which a number of slip bands and cracks appear on the fatigued specimen surface. Therefore, to properly apply the replica method for Ti-6Al-4V alloy, a special consideration is necessary.

To successfully observe fatigue cracks on the specimen surface, a notched specimen would be useful. This is due to the fact that the stress concentration will promote the initiation and the propagation of cracks intensively at the notched part. Shojima et al. [25] have conducted high- cycle fatigue tests for Ti-6Al-4V alloy in fully-reversed tension-compression by using specimens containing shallow circumferential notches and they succeeded in successive observation of the fatigue phenomena on the surface of notch root. However, the notches had a depth of 200 μm and a notch-radius of either 20 μm or 100 μm. The stress concentration factors were approximately 8 and 4, respectively, and this specimen can hardly be regarded as a smooth specimen. In addition, the deep sharp notch of this level has such a problem that it requires a skilled task to take replicas at the bottom of a deep notch. In this study, a specimen with a very small shallow notch of 20 μm in depth and 20 μm in radius was also used, but cracks at the notch root could not be observed in the high-cycle fatigue regime. This may be because the root of such a small notch rarely encounters the grains favorable to crack initiation due to the statistical factor mentioned above. To solve the above problems, in this study, a trial fatigue test for a specimen with a shallow blunt notch was performed in rotating bending. It is expected that the steep stress gradient generated by bending of small-diameter specimen would enhance the likelihood of crack initiation on the surface.

The material investigated was hot-rolled Ti-6Al-4V alloy with an equiaxial α+β microstructure. A small shallow notch of 50 μm depth and 100 μm radius was circumferentially introduced into the round-bar specimens of 3.4 mm diameter. The stress concentration factor was about 2.4. The specimen surface was electro-polished after notching to remove residual stress and to expose microstructure. Fatigue test was performed at a stress ratio of –1 by a rotating bending testing machine.

Figure 16 shows the fatigue process observed successively with the gold-deposited replicas. The microstructure at notch root and the propagation behavior of cracks are clearly shown. It is seen that crack initiation took place between the numbers of cycles of 1.6×10^4 and 3.0×10^4, and thereafter the crack grew by coalescing with other cracks. These cracks initiated and propagated within α-phase or along the interface of α- and β-phases. Figure 16 b) shows the magnified micrographs at A and B in Figure 16 a). It was observed under an optical microscope that the crack just initiated was a hairline crack and there seemed to be no slip bands at the crack initiation site. The high-resolution microscopy such as SEM or TEM may be useful to investigate the mechanism of crack initiation in more detail.

The above result is obtained from notched specimen and is not totally the same as that of smooth specimen. However, a lot of information relevant to crack initiation and early propagation can be obtained through this experiment. In essence, by increasing root radius and decreasing notch depth, the mechanical condition on the surface at notch root becomes similar to that on the surface of smooth specimen. It is expected that a series of experiments with a variety of notch geometries would reveal the fatigue mechanism of Ti-6Al-4V alloy that has not been well understood so far.

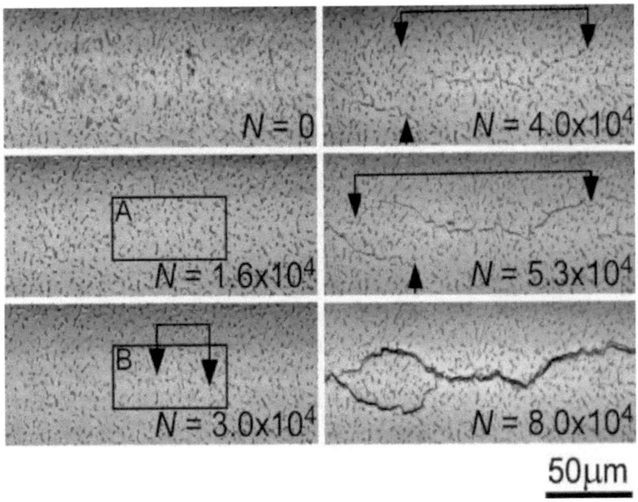

a) Change in surface state of notch-root

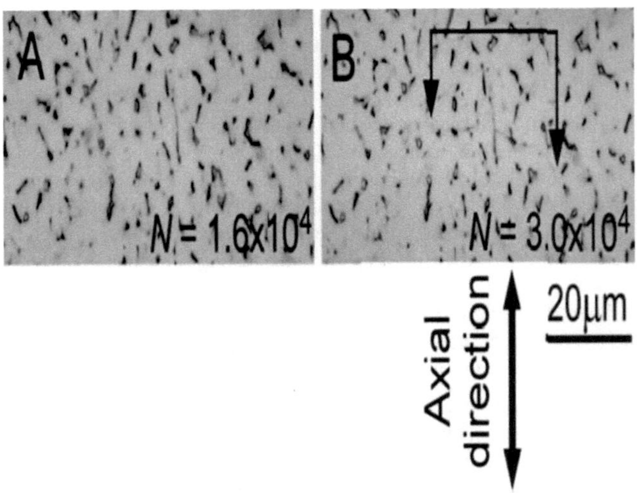

b) Magnified micrographs at A and B in a)

Figure 16. Successive observation of fatigue process at notch-root by the replica method; notch-radius = 100 μm, notch-depth = 50 μm, σ_a = 340 MPa (*R* = -1), and *N*f = 1.06×10^5

4. Concluding remarks

Fatigue behavior of ultra-fine grained Ti-6Al-4V alloy with grain size of 0.5 μm produced by Protium treatment was s ummarized mainly on the basis of the authors' recent experimental findings. The higher fatigue strength of the ultra-fine grained Ti-6Al-4V alloy was brought by delay of crack initiation in fatigue process. Fatigue crack initiation and propagation behavior of the ultra-fine grained Ti-6Al-4V alloy was characterized by fracture surface morphology such asgranular fracture. Striation and striation like fracture surface were very rarely observed on crack propagation area in air and in 3% NaCl aqueous solution. Corrosion fatigue crack with tiny corrosion pit was observed on specimen surface. Further investigations by use of the replica method are expected to reveal a mechanism on fatigue crack initiation and propagation in ultra- fine grained Ti-6Al-4V alloy in air and in corrosive environment such as 3% NaCl aqueous solution.

Acknowledgement

The authors would like to thank Associate Professor Keiji Yanase of Fukuoka University for valuable suggestions in preparing the manuscript.

5. References

1. H. Yajima, M. Yamamoto, T. Ishikawa and T. Fukui, Trans. The West-Japan Soc. Naval Architects, 94, 183-193 (1997).

2. H. Kimura, Y. Akiniwa, K. Tanaka, J. Kondo and T. Ishikawa, J. **Soc. Mat. Sci.,** Japan, 51, 795-800 (2002)

3. H. Kimura, Y. Akiniwa, K. Tanaka, J. Kondo and T. Ishikawa, **J. Soc. Mat. Sci., Japan**, 51, 801-807 (2002)

4. Y. Matsumura, K. Kishi, Y. Ishibashi and R. Ebara, **J. Soc. Mat. Sci.**, Japan, 57, 134-139 (2008)

5. Y. Yoshimura, K. Kimura, M. Hayashi, M. Ishii, T. Hanamura and J. Takamura, **J. Japan Inst. Metals**, 56, 1352-1359 (1992)

6. Y. Yoshimura, Y. Fujii, S. Wada, A. Murakami and K. Kimura, **J. Japan Inst. Metals**, 61, 83- 89 (1997)

7. T. Nakaiso, M. Tanada, R. Ebara , J. Nakahigashi, H. Yoshimura and K. Yamano, Proc. of the Fatigue, FT172 (8 pages) (2006).

8. L. R. Saitova, H. W. Höppel, M. Göken, I. P. Semenova, G. I. Raab and R. Z. Valiev, **Materials Science Engineering,** A503 145-147 (2009).

9. L. R. Saitova, H. W. Höppel, M. Göken, I. P. Semenova and R. Z. Valiev**, Intern. J. Fatigue**, 31, 322-331 (2009)

10. R. Ebara, K. Maeda, K. Kadaowaki, A. Murata, M. Miyoshi, T. Funakoshi, T. Ohtsu, J. Nakahigashi, Proc. of the grant workshop on New Methods of Damage and Failure Analysis of Structural Parts, Technical Univ. of Ostrava, 229-235 (2012)

11. A. Sugeta, Y. Uematsu, M. Yasuda and M. Jono, J. **Soc. Mat. Sci., Japa**n, 55, 545-549 (2006)

12. H. Mughrabi and H. W. Höppel**, Intern. J. Fatigue**, 32, 1413-1427 (2010)

13. J. Nakahigashi and H. Yoshimura, Transactions of JSME, 66, 2247-2253 (2000)

14. J. Nakahigashi and H. Yoshimura, **J. Alloys and Compounds**, 330-332, 384-388 (2002)

15. J. Nakahigashi and H. Yoshimura, **Materials Transactions**, 43, 2768-2772 (2002)

16. H. Yoshimura and J. Nakahigashi, **Intern. J. Hydrogen Energy**, 27, 769-774 (2002)

17. T. Nakaiso, Master of Engineering Dissertation, Kagawa University (2005)

18. R. Ebara, Y. Ishibashi, T. Yamane and T. Yamada, Proc. of the Fall Meeting of the JIM, 3, S578 (1977).

19. H. Kitahara, K. Uchikado, J. Makino, N. Iida, M. Tsushida, N. Tsuji, S. Ando and H. Tonda, **Materials Transactions**, 49, 64-68 (2008)

20. A. Shanyavsky, T. Zaharova and Y. Potahenko, Proc. of the VHCF4, edit. By J. E. Allison, J. W. Jones, J. M. Larsen and R. O. Ritchie, TMS, 325-330, (2007)

21. Y. Uto, R. Ebara, A. Hiromoto and K. Hatano, J. Japan Soc. Strength and Fracture of Materials,19, 1-8 (1984)

22. S. G. Ivanova, R. R. Biederman, and R. D. Sisson Jr., **J. Mater. Eng. Perform.**, 11, 226-231 (2002)

23. H. Oguma, T. Nakamura, S. Yokoyama and T. Noguchi, J. **Soc. Mat. Sci.,** Japan, 70, 1116- 1123 (2004)

24. V. Sinha, M. J. Mills, J. C. Williams and J. E. Spowart, **Metal. Mater. Trans**. A, 37, 1507- 1518 (2006)

25. K. Shojima, S. Weldle, S. Okazaki, M. Endo, D. Eifler and F. Balle, **Mater. Sci. Forum**, 750, 232-235 (2013)

FATIGUE of MATERIALS III
ADVANCES and EMERGENCES in UNDERSTANDING

Composites

Fatigue of Materials III: Advances and Emergences in Understanding
Edited by: T.S. Srivatsan, M. Ashraf Imam, and R. Srinivasan
TMS (The Minerals, Metals & Materials Society), 2014

QUASI-STATIC, FATIGUE AND FRACTURE BEHAVIOR OF ALUMINUM ALLOY COMPOSITE USED IN BRAKE DRUMS

K. Manigandan[1], T.S. Srivatsan[1], T. Quick[2]

1. *Department of Mechanical Engineering*
The University of Akron
Akron, Ohio 44325-3903, USA
E-Mail: tsrivatsan@uakron.edu

2. *Department of Geology*
The University of Akron
Akron, Ohio 44325, USA

Abstract

In this paper the results of a study aimed at investigating and understanding the quasi-static, cyclic fatigue properties and final fracture behavior of an aluminum alloy reinforced with particulates of ceramic, a viable candidate for use in brake drums of emerging automobiles, is highlighted. The processing treatment used to engineer the aluminum alloy composite is detailed. Specimens of both the aluminum composite and the preform that was used to engineer the composite were deformed in both tension and cyclic fatigue over a range of maximum stress in the room temperature, laboratory air environment and the number of cycles-to-failure was recorded. The stress-fatigue life response of the aluminum alloy composite is compared with the preform. Both the quasi-static and cyclic fatigue fracture surfaces of the deformed and failed samples were comprehensively examined in a scanning electron microscope to reveal the macroscopic fracture mode and to concurrently characterize the intrinsic features on the fracture surfaces.

Key words: aluminum alloy MMC, preform, microstructure, tensile, fatigue, fracture

1. Introduction

Metal matrix composites [MMCs] can safely and easily be categorized as composite material that comprises of at least two constituents, one being the primary metal, which incidentally is the matrix and the other being either a different secondary metal or another material, such as ceramics in the form of particulates, whiskers, short fibers or long-fibers [1-3]. When compared one-on-one with the monolithic counterpart, which comprises essentially of a single constituent, the metal-matrix composite [MMC] is known to offer a higher strength-to-density ratio, a higher stiffness-to-density ratio, better fatigue resistance and enhanced strength at elevated temperatures. The family of MMCs is also known to offer improved wear resistance when compared one-on-one to the monolithic counterpart [1-6]. As such, since the early 1980s, in the time period spanning the last three decades, MMCs have grown both in significance and strength to find increasing use for those applications that demand wear resistance and strength at elevated temperatures, such as brakes and several other vehicle applications besides aerospace-related applications [7, 8].

A MMC is often produced by augmenting the primary metal with the secondary metal or another material, which plays the role of reinforcement. The materials often used for the primary metal or matrix and the reinforcing material are chosen with the primary objective of achieving the best combination of physical and mechanical properties. During the last three decades, numerous combinations of metals and reinforcing materials have been tried and eventually emerged for practical use in both short-term and even long-term applications. A few examples of the primary metal or matrix that have been chosen and used are aluminum, magnesium, titanium, copper, zinc, the superalloys, and even the intermetallics. Examples of the reinforcing material range from (i) particulates of the ceramic to include boron carbide, silicon carbide, alumina, graphite, to (ii) continuous fibers, and (iii) discontinuous fibers or whiskers of the same selection [5-10].

One method that has been developed, tried and eventually emerged as an economically viable technique for the production of a MMC is to impregnate or infiltrate a preform of the reinforcing material with the primary metal. The preform is often a fabric or prearranged fibrous configuration of the reinforcing material that is produced and kept ready prior to infiltration.

Many of the prevailing preforms suffer from fatigue and eventual failure and often do not exhibit uniform strength in three dimensions. Additionally, most of the existing preforms that are favored for use typically comprise of small reinforcing materials, which enable consistent mixing during formation of the preform. However, consistent mixing often becomes difficult when size of the reinforcing material decreases. Smaller reinforcing materials often tend to limit both the mechanical and physical properties, such as wear resistance, fatigue resistance and density of the preform and in turn the resultant metal matrix composite.

As of the prevailing time period many of the existing preforms exhibit inconsistent density as a direct consequence of inconsistent mixing. Since consistent mixing essentially involves both a dispersion of the reinforcing materials through the metal matrix and uniform distribution of the reinforcing material through the preform so as to minimize their agglomeration. Inconsistent mixing often results in non-uniform distribution of the reinforcing materials in existing preforms and contributes to the preform having inconsistent density, physical properties, and inferior mechanical performance. Preforms having inconsistent density tend to wear readily and are therefore not useful for those applications that require good wear resistance. Existing preforms also suffer from weak points caused by entanglement and agglomeration of the reinforcing materials, which can be fine particulates of the chosen ceramic. Such preforms that suffer from the conjoint influence of entanglement and agglomeration of the reinforcing materials do not exhibit sufficient strength and stiffness. The resultant MMC that was formed from preforms that comprised of either entanglement or agglomeration or both, would often suffer from inadequate metal infiltration due to blockages caused by entanglement and agglomeration of the reinforcing materials.

The large volume of experimental research studies on discontinuously-reinforced light alloy based metal-matrix composites (DRMMCs), in the areas of cyclic deformation, crack initiation, crack propagation and fracture behavior, have been essentially on systems based on the non-ferrous metal matrices, such as the aluminum alloys [11-16]. Extensive use of DRA MMCs in performance-critical components often involves repeated loading, and therefore the fatigue response and resultant failure and fracture characteristics are of need and interest. Few of the earlier studies on mechanical behavior of aluminum alloy metal matrix composites have found complex yet synergistic relationships to exist between the cyclic fatigue properties and final fracture characteristics of a discontinuous particulate-reinforced aluminum alloy-based MMC [17-26].

In recent years, several related studies have focused on developing an understanding of the influence of the reinforcement, either particulates or fibers, on matrix microstructure and resultant influence of composite microstructure on both stress-controlled and strain controlled fatigue behavior of MMCs [18-26]. Complex relationship does exist between the fatigue properties and fracture characteristics of a reinforced aluminum alloy based MMC. These often include the following:

(i) The intrinsic properties of the aluminum alloy (composition, aging condition and microstructure) chosen.
(ii) Intrinsic properties of the reinforcing phase (composition, nature, morphology, size and volume fraction).
(iii) Influence of secondary processing on microstructure, and
(iv) Influence of test parameters, such as, (a) nature, (b) type, (c) magnitude, and even (d) severity of loading, on overall mechanical performance.

With this underlying background, the motivation for this assigned research study was to investigate and compare the microstructure, tensile properties, cyclic fatigue and fracture behavior of the two constituents of a novel brake drum, i.e. aluminum alloy composite and the preform that was used to make the composite, for use in emerging automobile-related applications. The tensile properties, cyclic fatigue and fracture response of these two constituents of the brake drum are discussed in light of mutually interactive influences of: (i) intrinsic microstructural effects, (ii) nature of loading, (iii) magnitude and/or severity of the stress amplitude, and (iv) deformation characteristics of the constituents.

2. Materials

The brake drum used in this specific investigation was provided by Bendix Corporation (Elyria, Ohio, USA). The MMC on the brake drum was prepared by infiltrating the molten aluminum alloy, a cast aluminum alloy, into a preform consisting of a mixture of ceramic particulates and short fibers. The ceramic fibers are included in the preform to facilitate reduction in density while concurrently enhancing metal infiltration and optimizing strength of the preform

3. Processing of the Aluminum alloy Composite:

In this study the metal matrix composite chosen contains a preform that was formed from a composition containing both ceramic particulates and ceramic fibers. The ceramic fibers had an aspect ratio greater than 3:1 and were randomly oriented in three dimensions. The base metal i.e., metal matrix, is allowed to impregnate through the preform resulting in the formation of the metal matrix composite. The resultant MMC was formed by extruding the composition through a multi-screw extruder to form an extrudate. The multi screw extruder tends to randomly orient the ceramic fibers in three dimensions as the composition is extruded through the multi-screw extruder. The method also includes the following [27]:

(a) Forming the extrudate to a desired configuration for purpose of defining the preform.
(b) Drying the preform, and
(c) Heating the preform to remove the organic binder while concurrently forming an open cell matrix.

Heating of the preform also aids in strengthening the preform so as to facilitate ease in its positioning within a portion of the cavity in the mold. The matrix material, i.e. metal, is heated above its melting point so as to form a molten state and then injecting the molten metal into the cavity of the mold under pressure so as to infiltrate the preform with the molten metal. Subsequent to complete infiltration the molten metal is cooled resulting in a metal matrix composite, which is eventually removed from the mold.

The ability of using a multi-screw extruder fur purpose of mixing composition of the preform ensures consistent mixing and a near uniform distribution of the reinforcing materials so as to prevent blockages while concurrently ensuring proper metal infiltration. The resultant metal matrix composite has randomly oriented ceramic fibers having an aspect ratio greater than 3:1. This method results in the production of a lightweight ceramic preform that meets or exceeds the strength requirements for those applications that require good resistance. Furthermore, the engineered metal matrix composite does not require any other type of external reinforcing element. Details pertaining to processing technique for producing the MMC used in this study can be found elsewhere [27].

4 Experimental Techniques

4.1 Initial Microstructure Characterization

Samples were cut from both the base metal and the MMC and prepared by standard metallographic procedures for observation in an optical microscope. The morphology of the reinforcing phases, i.e., both the particulates and fibers, and their distribution through the aluminum alloy metal matrix and other observable microstructural features were carefully examined in an optical microscope and photographed using a bright-field illumination technique.

4.2 Specimen Preparation

Smooth cylindrical test specimens (3.175 mm gage diameter and 12.5 mm gage length) were machined from blanks of both the aluminum composite and the preform using a diamond-tipped cutting tool. The test specimens conformed to specifications outlined in ASTM Standard E606 [28]. The length-to-diameter ratio of the mechanical test specimen was chosen so as to ensure that it would not buckle under fully reversed stress-amplitude ($R_\sigma = -1$)-controlled cyclic deformation. To minimize the effects of both surface irregularities and finish, final surface preparation was achieved by mechanically polishing the gage section of all test specimens, prepared from both the MMC and the preform sections of the brake drum, using progressively finer grades of silicon carbide impregnated emery paper, to remove any and all of the circumferential scratches and surface machining marks.

.

4.3 Mechanical Testing

All mechanical tests were performed on a fully automated, closed-loop, servo hydraulic structural test machine (INSTRON: Model 8500 Plus) equipped with a 100 KN load cell. A precision alignment was given to the load train prior to the initiation of testing so as to eliminate and/or minimize grip-induced bending. Tensile tests were carried out at a strain rate of 10^{-4}/s. In view of the anticipated limited ambient temperature ductility of both the aluminum composite and the preform, cyclic stress amplitude-controlled high-cycle fatigue tests were conducted at ambient temperature (T = 25°C) over a range of stress amplitudes. Cyclic fatigue tests were conducted in the axial total stress-amplitude control mode at a constant cyclic frequency of 5 Hz and at stress ratio [minimum stress (σ_{min}) / maximum stress (σ_{max})] of 0.1. The number of cycles to cause complete failure by separation or breakage is taken as fatigue life (N_f).

4.4 Failure-Damage Analysis

Fracture surfaces of the tensile and cyclically deformed and failed fatigue test specimens were comprehensively examined in a JEOL scanning electron microscope (SEM) to determine the following:

(i) Macroscopic fracture mode, and

(ii) Characterize the fine-scale topography and microscopic mechanisms governing final fracture.

Samples for SEM observation were obtained from the failed specimens by sectioning parallel to the fracture surface.

5. Results and Discussion

5.1 Undeformed Microstructure

The optical micrographs illustrating the microstructure of both the aluminum alloy composite and the preform, i.e. the two key ingredients of the as-provided brake drum, are shown in **Figure 1** and **Figure 2.** The preform essentially revealed a duplex microstructure consisting of light colored regions, identified to be the reinforcing phase, i.e., fine ceramic particulates, dispersed randomly through the aluminum alloy metal matrix. At frequent intervals an agglomeration or clustering of the reinforcing ceramic particulates was observed resulting in pockets of reinforcement-rich and reinforcement-depleted regions. The reinforcing fibers, if any, could not be easily resolved and observed at the allowable magnifications of the light microscope. The observable reinforcing ceramic particulates were of varying size and near elliptical in shape as shown in **Figure 1**. In this research study, no attempt was made to determine the particle size distribution of the reinforcing ceramic particulates in the preform.

The light optical micrograph of the aluminum alloy composite portion of the brake drum revealed intrinsic features quite typical of a reinforced aluminum alloy (**Figure 2**) with ceramic particulates of very fine and small in size dispersed through a eutectic matrix.

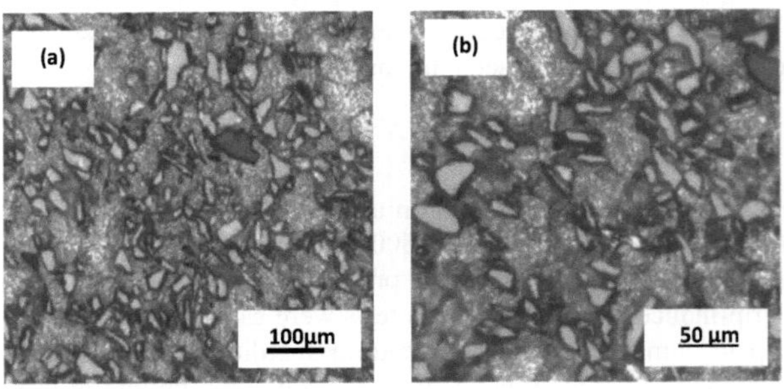

Figure 1. **Light optical micrographs of the preform portion of the aluminum brake drum at two observable magnifications showing the nature, morphology and distribution of the ceramic particulates in the preform.**

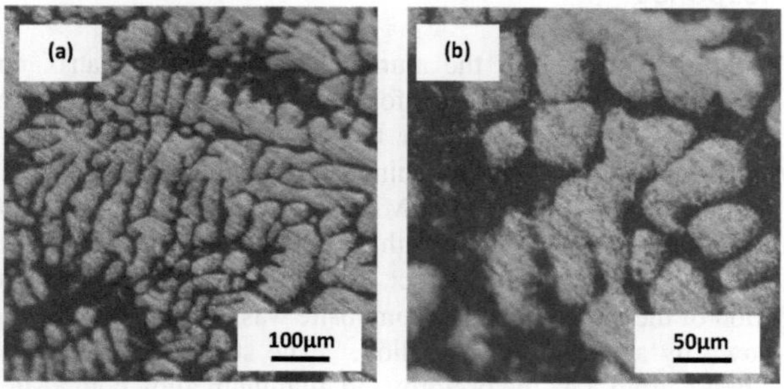

Figure 2. Light optical micrographs of the aluminum alloy metal matrix composite portion of the brake drum, at two different magnifications, showing a eutectic microstructure quite typical of a cast aluminum alloy that was inlaid with dispersion of ceramic particles.

5.2 Hardness

Both macro-hardness and micro-hardness of the preform were noticeably higher than that of the aluminum alloy metal matrix composite. For both the preform and the aluminum alloy composite the micro-hardness value was observably higher than the macro-hardness value quantified by measurement of Rockwell hardness. The micro-hardness and macro-hardness values are summarized in **Table 1** and shown pictorially in the bar graph in **Figure 3.**

Table 1: Summary of macrohardness and microhardness values of the two constituents of the brake drum:
(a) Aluminum alloy metal matrix composite, and
(b) Ceramic preform

Material	Macro-Hardness (average)		Micro-Hardness (average)	
	HR_{15T}	MPa	H_V	MPa
Aluminum alloy MMC	78	416	195	631
Ceramic Preform	83	496	221	739

5.3 Tensile Response

The tensile properties of both the aluminum alloy metal matrix composite and the preform are summarized in **Table 2**. The preform had yield strength of 261 MPa and a tensile strength of 295 MPa. The observed higher tensile strength beyond yield provides for the occurrence of limited amount of strain hardening or strengthening beyond yield. Both yield strength (110 MPa) and tensile strength (112 MPa) of the aluminum composite was noticeably lower than that of the preform. The preform that is used to engineer the aluminum composite revealed low ductility as evidenced by 2.0 pct. Elongation over 12.5 mm gage length. The ductility or elongation of the aluminum alloy composite was low and only 0.4 pct. commensurate with properties shown by a cast aluminum alloy. The stress versus strain curve of the two constituents of the brake drum, i.e., the preform and aluminum alloy composite, are compared in **Figure 4**. Both the aluminum composite and the preform exhibited limited ductility, on a macroscopic scale, with fracture occurring on plane normal to the far-field stress axis.

Table 2: A compilation of the room temperature tensile properties of the two key constituents of the aluminum Brake Drum (Temperature T = 25°C)

Material	Elastic Modulus		Yield Strength		Ultimate Tensile Strength		Elong. [12.5 mm]
	Ksi	GPa	ksi	MPa	ksi	MPa	(%)
Ceramic Preform	9354.93	64.5	37.85	261.0	42.79	295.0	2.0
Aluminum Alloy MMC	9050.35	62.4	16.01	110.4	16.21	112.0	0.4

Tensile fracture behavior of the aluminum alloy composite was overall flat and normal to the applied far-field stress axis (**Figure 5a**) High magnification observation of the fracture surface revealed an overall microscopically rough fracture surface with the presence of voids of varying size, dimples of varying size and shape, and even an array of very fine microscopic cracks. High magnification observation of **Figure 5(b)** revealed the morphology and flow of both the fine microscopic and macroscopic cracks (**Figure 5c**). The overload region of the tensile fracture surface was covered with features reminiscent of locally occurring ductile failure mechanisms **(Figure 5d)**.

Tensile fracture behavior of the ceramic preform is shown in **Figure 6.** As for the composite the overall morphology was flat and normal to the far field stress axis. Gradual high magnification observation of the fracture surface revealed it to be microscopically rough (**Figure 6b**). When the fracture surface was observed at higher allowable magnifications of the scanning electron microscope it revealed voids of varying size to be intermingled with shallow dimples and few very fine microscopic cracks; features reminiscent of locally occurring ductile and brittle failure mechanisms (**Figure 6c**). The region of overload when observed at higher magnifications revealed pockets of shallow dimples intermingled with very fine microscopic voids and few fine microscopic cracks indicative of the locally occurring failure mechanisms (**Figure 6d**).

Overall, macroscopic appearance was brittle but microscopically both the aluminum composite and the ceramic preform revealed features indicative of "locally" occurring brittle and ductile failure mechanisms.

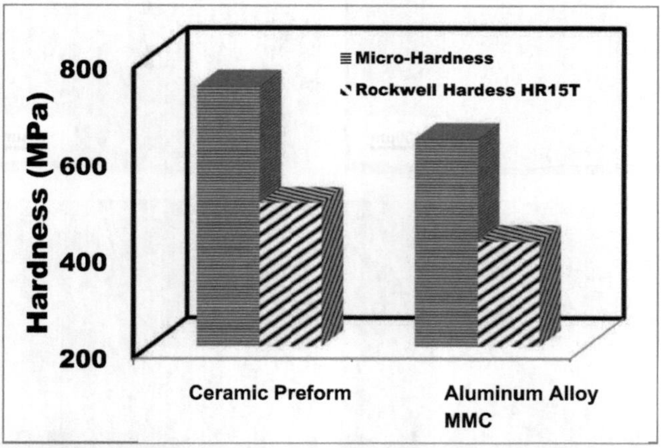

Figure 3. **Bar graph comparing the average value of microhardness (kg/mm^2) and macrohardness (Kg/mm^2) of the preform and aluminum alloy composite that form the very essence of the brake drum**

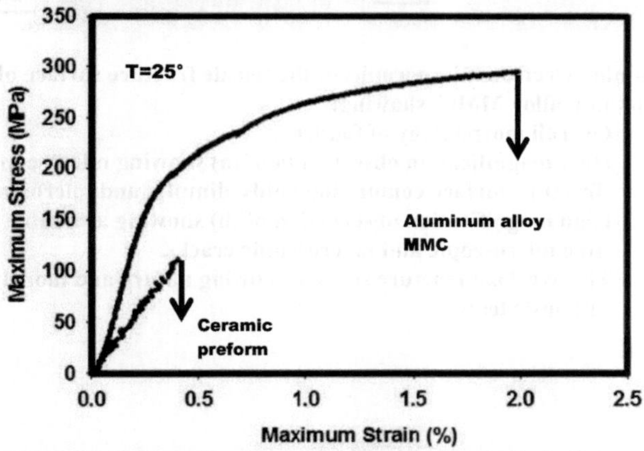

Figure 4. **Comparison of the engineering stress versus engineering strain curves of both the preform and aluminum alloy MMC the two key ingredients of the brake drum when deformed in uniaxial tension at room temperature (T = 25°C).**

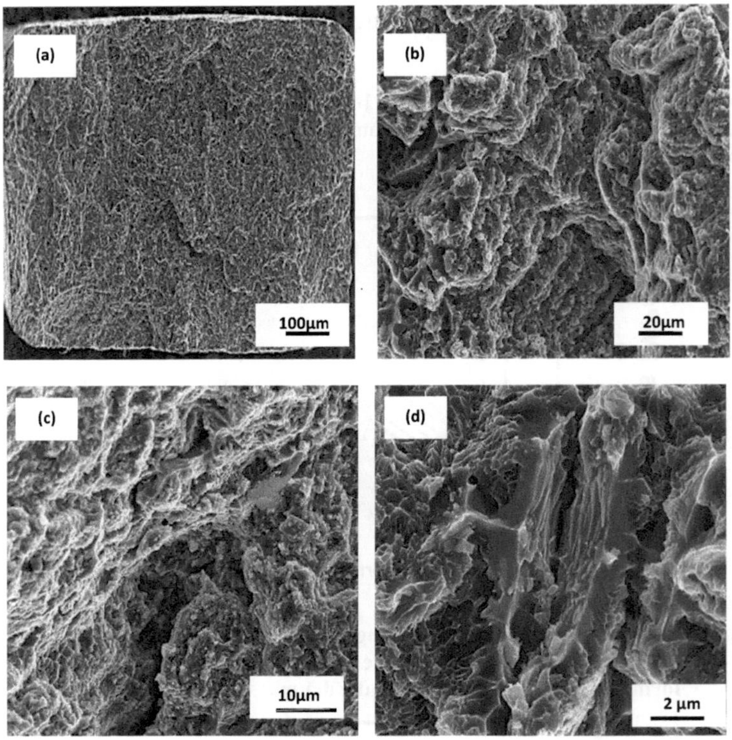

Figure 5. **Scanning electron micrographs of the tensile fracture surface of the aluminum alloy MMC, showing:**

(a) Overall morphology of failure

(b) High magnification observation of (a) showing microscopically rough fracture surface comprising voids, dimples and microscopic cracks.

(c) High magnification observation of (b) showing a sizeable number of fine microscopic and macroscopic cracks.

(d) The overload fracture surface showing nature and morphology of the intrinsic features.

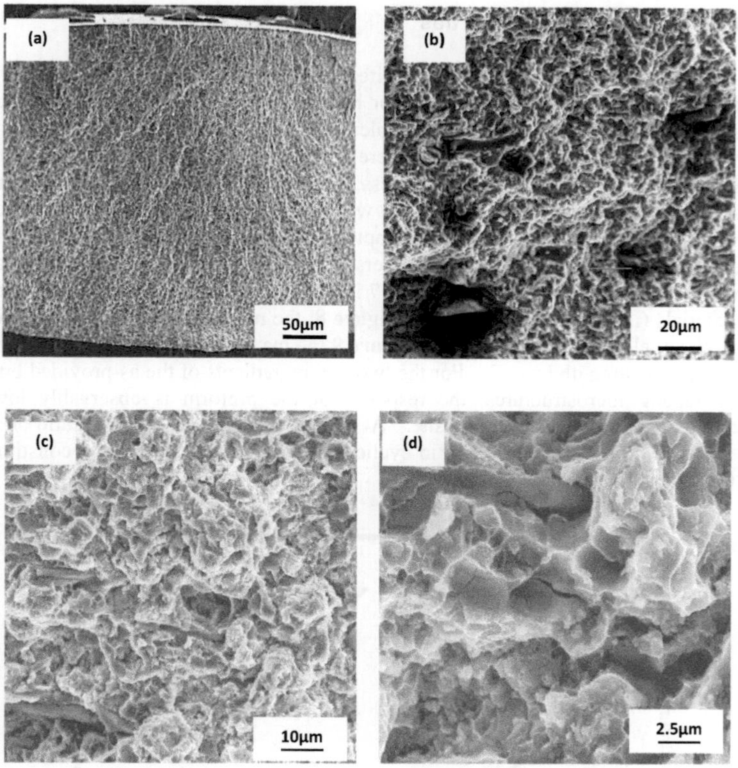

Figure 6. **Scanning electron micrograph of the tensile fracture surface of the ceramic preform showing the following:**

(a) Overall morphology of failure

(b) High magnification observation of (a) showing microscopically rough fracture surface and the intrinsic features.

(c) Voids of varying size intermingled with shallow dimples and very fine microscopic cracks on the transgranular fracture surface.

(d) Shallow dimples, fine microscopic voids and fine microscopic cracks covering the overload fracture surface indicative of ductile and brittle mechanism.

5.4 High Cycle Fatigue Properties

The maximum stress versus fatigue life response followed the trend shown by most metals and their composite counterparts. For both the aluminum composite and the preform with a decrease in stress amplitude a noticeable increase in fatigue life was evident. Both the aluminum alloy composite and the preform were cyclically deformed over a range of maximum stress, taken as specified ratios of the yield stress. At near equivalent values of maximum stress, the stress-fatigue life response of the preform was noticeably lower than that of the aluminum alloy composite **(Figure 7).** In an attempt to rationalize the specific role of intrinsic microstructural features in governing stress versus fatigue life response of both the aluminum MMC and the preform the test data in **Figure 7** is replotted to take into account the strengths of the two materials (i.e., microstructures). In **Figure 8**, the maximum stress has been normalized by the uniaxial, yield stress (σ_{YS}) while in **Figure 9** the maximum stress has been normalized by the ultimate tensile strength [σ_{UTS}]. For the two key ingredients of the as-provided brake drum, with the resultant microstructures, the response of the preform is observably lower when compared to the aluminum alloy composite. At near equivalent values of the ratio of $\sigma_{maximum}/\sigma_{YS}$ and $\sigma_{maximum}/\sigma_{UTS}$ the difference in cyclic fatigue life between the two constituents is as high as 100 percent.

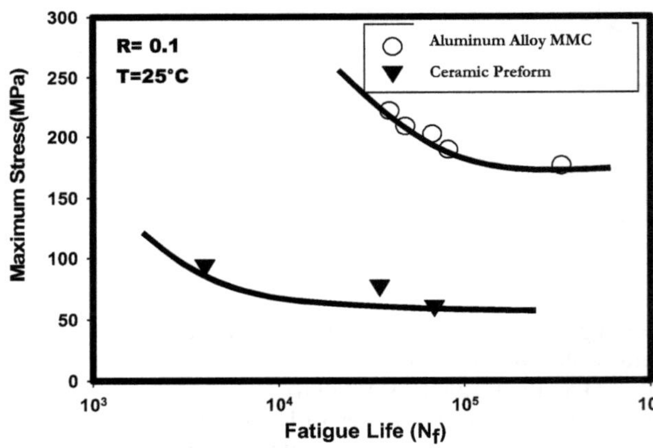

Figure 7. **A comparison of the maximum stress ($\sigma_{maximum}$) versus fatigue life (N_f) response of two ingredients of the brake drum, i.e., preform region and aluminum alloy MMC, when cyclically deformed at 25°C.**

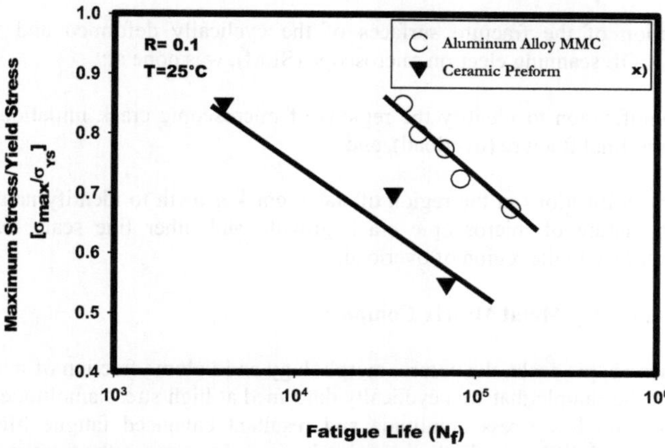

Figure 8. **Variation of ratio of maximum stress to yield stress (σ_{max}/σ_{YS}) with cyclic fatigue life (N_f) for two essential ingredients of the brake drum, i.e., aluminum alloy MMC and ceramic preform, when cyclically deformed at 25°C.**

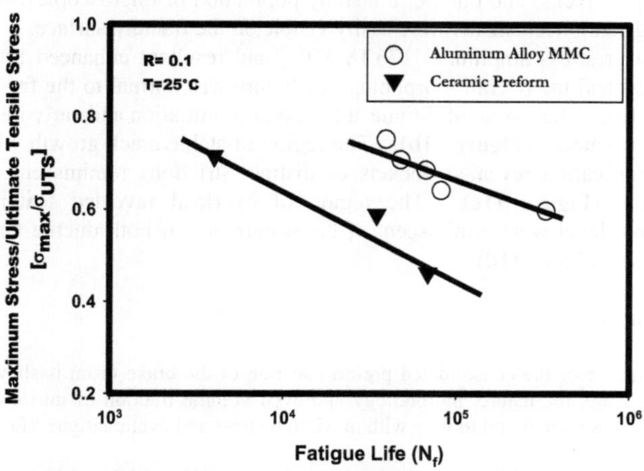

Figure 9. **A comparison of the influence of ratio of maximum stress (MPa) to tensile stress on cyclic fatigue life (N_f) of the two ingredients of the brake drum, i.e., ceramic preform and aluminum alloy MMC, when cyclically deformed at 25° C.**

5.5 Cyclic Fracture Behavior

Examination of the fracture surfaces of the cyclically deformed and failed fatigue specimens, in a JEOL scanning electron microscope (SEM), was done at:

(a) Low magnification to identify the regions of microscopic crack initiation, stable crack growth and final fracture (overload), and

(b) Higher magnifications in the region of stable crack growth to identify nature of damage initiation, nature of microscopic crack growth, and other fine scale features on the fracture surface in the region of overload.

(A) Aluminum Alloy Metal Matrix Composite

On a macroscopic scale, the nature, morphology and volume fraction of intrinsic features on the surface of the sample that was cyclically deformed at high stress amplitude, short fatigue life (**Figure 10**), and low stress amplitude and resultant enhanced fatigue life (**Figure 11**) revealed only marginal difference in the nature of intrinsic features on the fracture surface at the fine microscopic level.

(i) At a higher stress amplitude [221.8 MPa] and concomitant short fatigue life [39,544 cycles] features reminiscent of both brittle (combination of both fine microscopic and macroscopic cracks) and ductile (a healthy population of microscopic voids and dimples of varying size) mechanisms were easily visible on the fracture surface.

(ii) At the lower stress amplitude [176.18 MPa] and resultant enhanced fatigue life [Nf = 332,421 cycles] the overall morphology of failure was normal to the far-field stress axis (**Figure 11a**). The region of fatigue at the point of initiation and early microscopic crack growth was smooth (**Figure 11b**). The region of stable crack growth when observed at high magnification revealed pockets of distinct striations reminiscent of microplastic deformation (**Figure 11c**). The region of overload revealed features at the fine microscopic level were reminiscent of the occurrence of both ductile and brittle failure mechanisms (**Figure 11d**).

(B) The Preform

Fracture surface of the consolidated preform section of the brake drum is shown in **Figure 12**. On a macroscopic scale, the nature, morphology and even volume fraction of intrinsic features on the fatigue fracture surfaces were found to vary with maximum stress and cyclic fatigue life.

(i) For test sample of the preform that was cyclically deformed at a high maximum stress [77.7 MPa] and resultant short fatigue life of 35,014 cycles the overall morphology was essentially normal to the far-field stress axis (**Figure 12a**). High magnification observation revealed the region of crack initiation and early microscopic crack growth to be essentially smooth and distinct (**Figure 12 b**). Pockets of very shallow striations were observed in the region of stable crack growth indicative of the occurrence of micro-plastic deformation at the local level (**Figure 12c**). The region of overload revealed an observable population of fine microscopic voids intermingled with shallow dimples of varying size and isolated fine microscopic cracks, features reminiscent of the occurence of both ductile and brittle failure mechanisms at the fine microscopic level (**Figure 12d**).

(iii) Test specimen of the preform that was cyclically deformed at a low maximum stress of 61.05 MPa with a resultant fatigue life of 68,974 cycle's revealed overall morphology of failure to be flat and normal to the far-field stress axis (**Figure 13a**). The region of early microscopic crack growth was microscopically rough **(Figure 13 b**). The overload region of the fracture surface was rough and covered with features representative of both ductile and brittle failure mechanisms occurring at the fine microscopic level (**Figure 13c**).

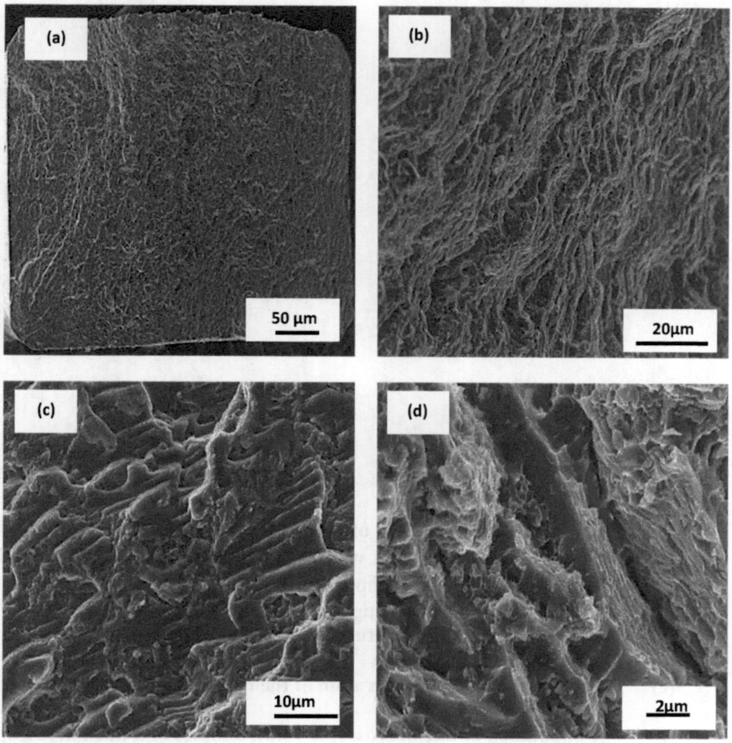

Figure 10. **Scanning electron micrograph of the high cycle fatigue fracture surface of the aluminum alloy MMC that was cyclically deformed at stress amplitude of 221.85 MPa with a fatigue life (N_f) of 39,544 cycles, showing:**

(a) Overall morphology of failure.

(b) High magnification observation of (a) showing the region of crack initiation and early microscopic crack growth.

(c) The region of stable crack growth showing fine striation like features reminiscent of local microplastic deformation.

(d) The region of overload showing shallow dimples, microscopic cracks arising from void coalescence; reminiscent of ductile and brittle mechanisms.

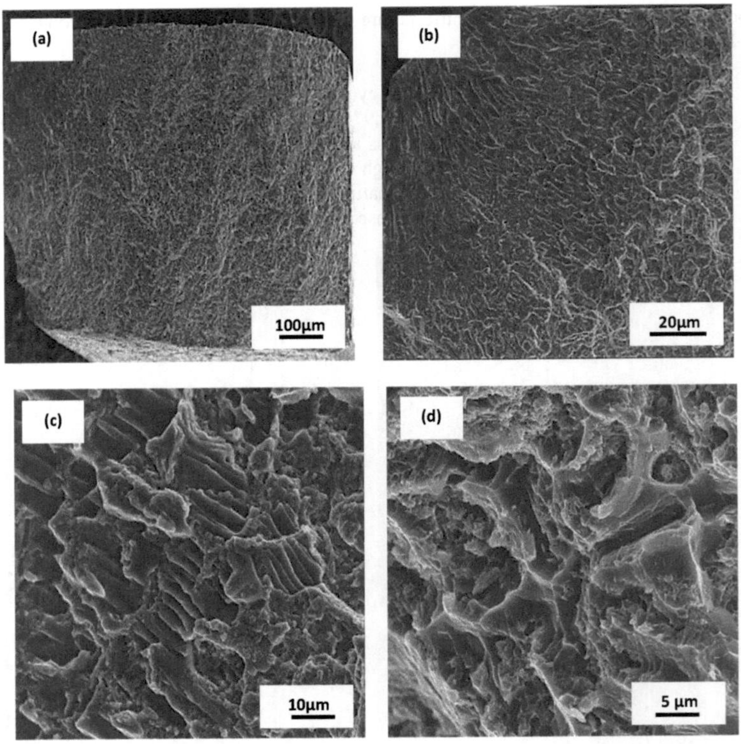

Figure 11. **Scanning electron micrograph of the high cycle fatigue fracture surface of the aluminum alloy MMC that was cyclically deformed at 176.18 MPa, fatigue life of 332,421 cycles, showing:**

(a) **Overall morphology of fatigue failure**

(b) **The region of fatigue fracture surface at the point of initiation and early microscopic crack growth**

(c) **High magnification observation of the region of stable crack growth showing striations.**

(d) **A population of fine microscopic voids, dimples and fine microscopic cracks on the overload fracture surface; features reminiscent of locally ductile and brittle failure mechanisms.**

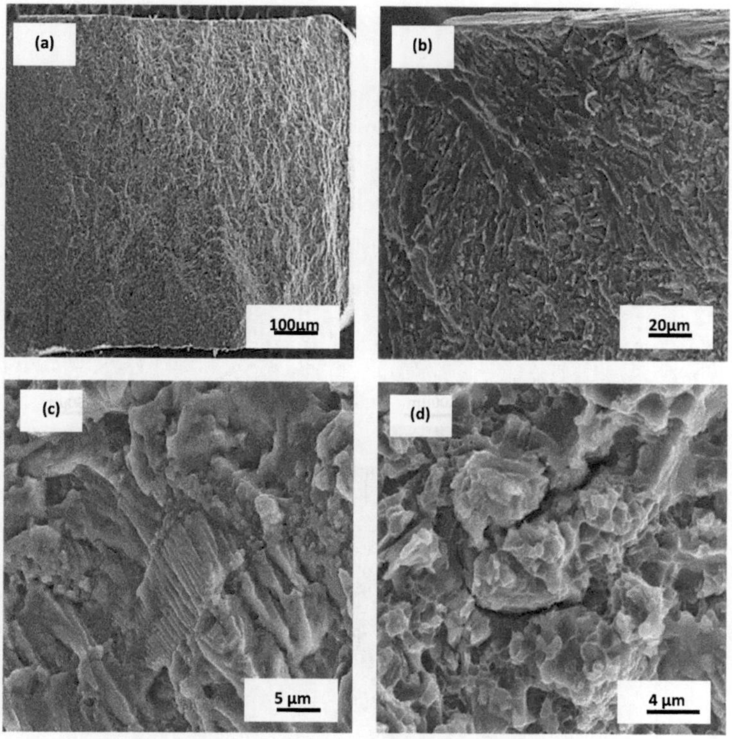

Figure 12. **Scanning electron micrographs of the surface of the test specimen of ceramic preform that was cyclically deformed at a maximum stress of 77.7 MPa, with fatigue life (N_f) of 35,014 cycles, showing:**

(a) **Overall morphology of fatigue failure**

(b) **High magnification observation of (a) showing microscopically the region of crack initiation and early microscopic crack growth.**

(c) **Pockets of shallow striations in the region of stable crack growth.**

(d) **High magnification observation of the overload region showing the features spanning isolated microscopic cracks, dimples of varying size, intermingled with isolated microscopic voids.**

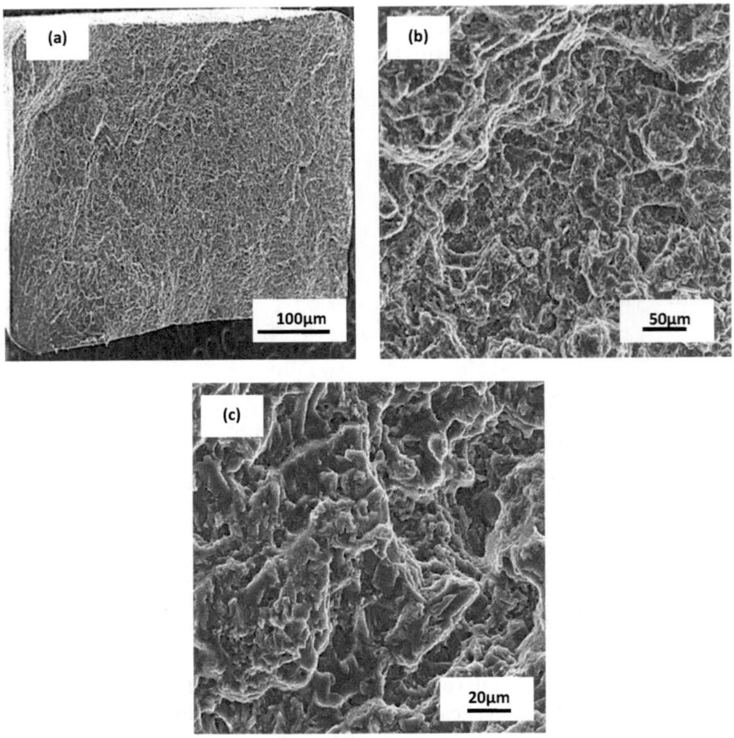

Figure 13. **Scanning electron micrographs of the surface of test specimen of the ceramic preform that was cyclically deformed at a maximum stress of 61.05 MPa, with fatigue life of 68,974 cycles, showing:**

(a) Overall morphology of failure

(b) High magnification observation of (a) in the region of early Microscopic crack growth showing rough fracture surface

(c) The key features covering the fracture surface in the region of overload failure.

6. Summary and Conclusions

Based on a study aimed at understanding the specific role and influence of ceramic reinforcements, both particulates and fine fibers, on microstructure, tensile response, stress amplitude-controlled high cycle fatigue and final fracture behavior of the two ingredients, i.e., preform and aluminum alloy composite, of a novel and emerging brake drum for use in brake drum of emerging automobiles, following are the key observations and findings:

1. The preform revealed a duplex microstructure consisting of light colored regions, identified to be the reinforcing ceramic particulates, dispersed through the aluminum alloy metal matrix. At frequent intervals an agglomeration of the reinforcing ceramic particulates was observed resulting in pockets of particle-rich and particle-depleted regions. The reinforcing particulates were of varying size and near elliptical in shape.

2. Light optical microstructure of the aluminum alloy metal matrix composite portion of the brake drum revealed intrinsic features quite typical of a cast aluminum alloy with pockets of varying size ceramic particulates dispersed through a eutectic matrix.

3. Both macrohardness and microhardness of the preform was noticeably higher than that of the resultant metal composite. For both the preform and the aluminum composite the microhardness value was observably higher than the macrohardness.

4. The preform had yield strength of 261 MPa. and a tensile strength of 295 MPa. The observed higher tensile strength beyond yield provides for limited amount of strain hardening or strengthening beyond yield. Further, the preform revealed low ductility as evidenced by elongation of only 2.0 pct. The ductility or elongation of the aluminum alloy composite was low and only 0.4 pct.

5. The stress – fatigue life response followed the trend shown by most metals and their composite counterparts. With a decrease in stress a noticeable increase in fatigue life was evident. At equivalent values of maximum stress the stress-fatigue life response of the preform was noticeably lower than the aluminum alloy composite.

6. On a macroscopic scale, the nature, morphology and volume fraction of the intrinsic features on the fatigue fracture surfaces were found to vary with maximum stress ($\sigma_{maximum}$) and cyclic fatigue life (N_f). For this preform section of the brake drum observation of pockets of striations in the regions of early microscopic crack growth and stable crack growth is indicative of the occurrence of micro-plastic deformation at the 'local' level. The region of overload of the preform revealed a population of fine microscopic voids intermingled with shallow dimples of varying size. This suggests the occurrence of ductile failure mechanisms at the fine microscopic level for the composite.

7. For the aluminum alloy metal matrix composite that was cyclically deformed at high stress, short fatigue life and low stress, long fatigue life, there was only a marginal difference in the nature of intrinsic features on the fracture surface at the fine microscopic level. At high stress amplitude and concomitant short fatigue life, features reminiscent of both brittle and ductile failure mechanisms were visible on the fracture surface.

However, at the lower stress amplitudes and enhanced fatigue life the features at the fine microscopic level were reminiscent of the occurrence of brittle failure mechanisms.

References

1. S. V. Nair, J. K. Tien and R. C. Bates: **International Metals Reviews**, Vol. 30, No. 6, 1985, pp. 285-297.

2. M. Taya, M and R. J. Arsenault: Metal Matrix Composites: Thermomechanical Behavior, Pergamon Press, Elmsford, New York , 1989.4.

3. T. S. Srivatsan and T. S. Sudarshan (editors): Rapid Solidification Technology: An Engineering Guide, Technomic Publishing Inc., Lancaster, PA, 1993, pp. 603-700.

4. M. Taya, and T.Mori, "Strengthening Mechanisms of Metal Matrix Composites," **Acta Metallurgica**, Vol. 35, pp. 155, 1987.

5. R.J. Arsenault, and R.K. Everett, (editors), Metal-Matrix Composites: Mechanisms and Properties, Academic Press, New York, USA, 1991.

6 J. E. Allison and J. W. Jones: "Fatigue behavior of discontionuously-reinforced metal-matrix composites," in **Fundamentals of metal matrix composites** (editors: S. Suresh, A. Mortensen and A. Needleman), Butterworth-Heinemann, 1993, pp. 269-296.

7. W. H. Hunt, Jr., C.R. Cook and R. R. Sawtell: "Cost Effective High-Performance Powder Metallurgy Aluminum Matrix Composites for automotive applications," 1991 Annual SAE Congress, Paper No. 910834, Detroit, Michigan, 1991.

8. W. H. Hunt, Jr.: "Cost-Effective high performance aluminum matrix composites for aerospace applications," International Conference on PM Aerospace Materials, MPR Publishing, Laussane, Switzerland, 1991, pp. 32-1 to 32-15.

9. J. J. Lewandowski, C. Liu and W. H. Hunt, Jr.: in Powder Metallurgy Composites (editors: P. Kumar, K. Vedula and A. M. Ritter), TMS: The Minerals, Metals and Materials Society, Warrendale, PA, 1989.

10. D. L. McDanels: **Metallurgical Transactions**, Vol. 16A, 1985, pp. 1105-1115.

11. M.J. Koczak, S.C. Khatri, J.E. Allison and M.G. Bader: in **Fundamentals of Metal Matrix Composites** (editors: S. Suresh, A. Mortensen and A. Needleman), Butterworth-Heinemann, Boston, MA., pp. 297-317, 1993.

12. J. K. Shang, and R. O. Ritchie: "Crack bridging by uncracked ligaments during fatigue crack growth *in silicon carbide reinforced aluminum alloy composite,"* **Metallurgical Transactions**, Vol. 20A, 1989, pp. 897-908.

13. J. K. Shang and R. O. Ritchie: "On the particle size dependence of fatigue crack propagation thresholds in SiC particulate-reinforced aluminum alloy composites," **Acta Metallurgica**, Vol. 37, 1989, pp. 2267-2278.

14. D. L. Davidson: "The growth of fatigue cracks through particulate silicon carbide reinforced aluminum alloys," **Engineering Fracture Mechanics,** Vol. 33, 1989, pp. 965-978.

15. Y. Sugimura and S. Suresh: "Effects of SiC content on fatigue crack growth in aluminum alloys reinforced with SiC particles," **Metallurgical Transactions**, Vol. 23A, 1992, PP. 2231-2242.

16. M. Manoharan and J. J. Lewandowski: **Acta Metallurgica**, Vol. 38, No. 3, 1990, pp. 489-499.

17. S. Kumai, J. E. King and J. F. Knott: "Short and long fatigue crack growth in a SiC reinforced aluminum alloy," **Fatigue and Fracture of Engineering Materials and Structures**, Vol. 13, 1990, pp. 511-520.

18. D. L. Davidson: "Fracture toughness of particulate metal-matrix composites," in Metal Matrix Composites: Mechanisms and Properties (edited by: R.K. Evertt and R. J. Arsenault), Academic Press, San Diego, CA 1991, pp. 217-230.

19. D. L. Davidson: "Fracture characteristics of Al-4 Pct. Mg mechanically alloyed with SiC," **Metallurgical Transactions**, Vol. 22A, 1991, pp. 97-110.

20. T.S. Srivatsan, K. Yamaguchi, and E.A. Starke, Jr., "Effect of Environment and Temperature on the Low Cycle Fatigue Behavior of Aluminum Alloy 2020," **Materials Science and Engineering**, Vol. 83, pp. 87-107, 1986.

21. T.S. Srivatsan, and E.J. Coyne, Jr., "Cyclic Stress Response and Deformation Behavior of Precipitation Hardened Aluminum-Lithium Alloys," **International Journal of Fatigue**, Vol. 8(4), pp. 201-210, 1986.

22. T. Christman and S. Suresh: **Materials Science and Engineering**, Vol. 102, 1988, pp. 211-220.

23. T.S. Srivatsan, and E.J. Coyne, Jr., "Micro mechanisms Governing the Fatigue Behavior of Aircraft Structural Alloys," **Materials Science and Technology,** Vol. 5, pp. 548-558, 1989.

24. T. S. Srivatsan, R. Auradkar, and A. Prakash, "Cyclic Stress Response and Cyclic Fracture Behavior of Silicon Carbide Reinforced Aluminum Metal Matrix Composite," **Engineering Fracture Mechanics,** Vol. 40, No. 2, 1991, pp. 277-295, 1991.

25. T. S. Srivatsan: "The low-cycle fatigue behavior of an aluminum alloy metal matrix composite," **International Journal of Fatigue**, Vol. 14, No. 3, 1992, pp. 355-366.

26. T. S. Srivatsan and R. Auradkar: "Effect of silicon carbide particulate on cyclic plastic strain response characteristics and fracture of aluminum alloy composites," **International Journal of Fatigue,** Vol. 14, No. 6, 1992, pp. 173-183.

27. United States Patent No. US 8,016,018 B2 Method of Manufacturing a Metal Matrix Composite, September 2011

28. ASTM E-606: "Stress Amplitude Controlled Fatigue of Metallic Materials," ASTM, American Society of Testing and Materials, Philadelphia, PA 1993.

29. T. S. Srivatsan and M. Al-Hajri: "The fatigue and final fracture behavior of SiC particle-reinforced 7034 aluminum matrix composites, **Composites, Part** B, Vol. 33, 2002, pp. 391-404.

Acknowledgements

This research effort was made possible through material (brake drum) and limited funding provided by **Bendix Commercial Vehicle Systems LLC** (Elyria, Ohio, USA) [Program Manager: Mr. S. Shaver; Program Director: Mr. M. Hamdan].

Fatigue of Materials III: Advances and Emergences in Understanding
Edited by: T.S. Srivatsan, M. Ashraf Imam, and R. Srinivasan
TMS (The Minerals, Metals & Materials Society), 2014

EFFECT OF COBALT CONTENT ON FATIGUE LIFETIMES AND SHORT FATIGUE CRACK GROWTH BEHAVIOR OF TUNGSTEN-COBALT CEMENTED CARBIDES

Hiroko Mikado[1, 2], Sotomi Ishihara[3], Noriyasu Oguma[1], Kenichi Masuda[1], Shingo Kawamura[2]

[1] **University of Toyama;**
3190 Gofuku, Toyama
Toyama PREF., 930-8555, **Japan**

[2] **YKK Corporation;**
200 Yoshida, Kurobe,
Toyama PREF., 938-8601, **Japan**

[3] **National Institute of Technology,**
Toyama College;
Toyama, Toyama PREF., 939-8630, **Japan**

Abstract

Fatigue lifetimes and crack growth behavior of the cemented carbides with different Co content were investigated. Effect of the Co content on the fatigue lifetimes is found to be minimal within the Co content, 13 –18 wt% tested in the present study. At the high *K*max region, the crack growth resistance for the WC-18 wt%Co is higher than that for the WC-13 wt% Co. In the lower *K*max region, the reverse trend, i.e., the crack growth resistance for the WC-13 wt% Co is higher than that for the WC-18 wt%Co, was expected, but this point remain obscure due to the lack of the experimental data.

Keywords: Fatigue, Crack Growth, Cemented Carbide, Co Content

1. Introduction

WC-Co cemented carbides have been widely used for cutting tools, die tools, and machine parts because of their excellent wear resistance, which is superior to that of tool steels [1]. To use them effectively without failure, it is important to understand their fatigue crack growth behavior as well as their fatigue lifetime. Although many studies [2, 3] have already been reported on the fatigue lifetime of WC-Co cemented carbides, a few investigations [4, 5] have focused on their fatigue crack growth behavior. So, further studies on the fatigue crack growth behavior of the WC-Co cemented carbides are therefore needed because fatigue data is absolutely lacking from the above viewpoint.

In the present study, rotating bending fatigue tests were conducted on two grades of fine grained WC-Co cemented carbides with different Co content. To evaluate their fatigue lifetimes and crack growth behavior, the plane and artificially notched specimens were used respectively.

2. Material and Experimental Procedures

2.1 Material and test specimens

Commercially available fine-grained WC-Co cemented carbides with different Co content (13 wt% Co and 18 wt% Co) were used as the test material. In this study, they are labeled as WC-13%Co and WC-18%Co, respectively. Tables I and II show the chemical compositions and mechanical properties of the WC-Co cemented carbides. Figure 1 shows the microstructures of the materials, which are composed of both WC grains and the Co phase. Average grain diameters of WC and particle size distributions of WC for each material are almost the same.

Table I. Chemical compositions of the test specimen (wt%).

Specimen	Co	Cr	W and C
WC-13%Co	13.0	0.51	Bal.
WC-18%Co	17.7	0.46	Bal.

Table II. Mechanical properties of the test specimen.

Specimen	Young modulus [GPa]	Fracture toughness [MPam$^{1/2}$]	Bending strength [MPa]	Vickers hardness [HV]
WC-13%Co	550	12.1	4,550	1,480
WC-18%Co	490	13.1	4,450	1,310

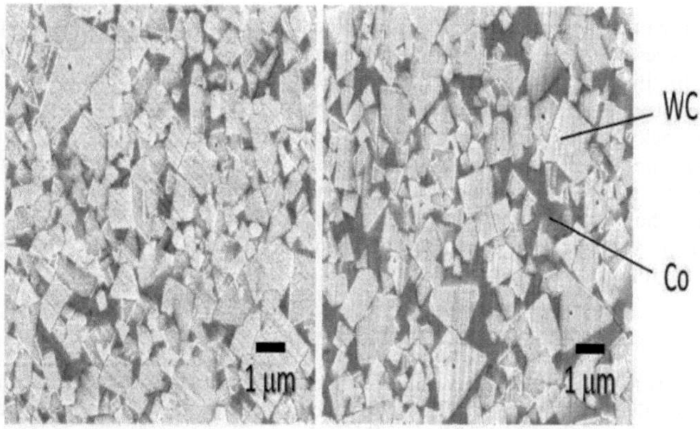

(a) WC-13%Co **(b) WC-18%Co**

Figure 1. Microstructures of WC-Co cemented carbide used in the present study. The average WC grain diameter was approximately 0.45μm.

Figure 2 shows the hourglass-shaped specimens used in the present study. Its minimum diameter was 3 mm. They were polished into a mirror-like finish with a surface roughness (*R*a) of less than 0.1 μm. Figure, and 2b 2a shows showsthenotched specimenthe. Forplane s the latter, the artificial notch with a length of 8 introduced into the center of the specimen by laser beam machining.

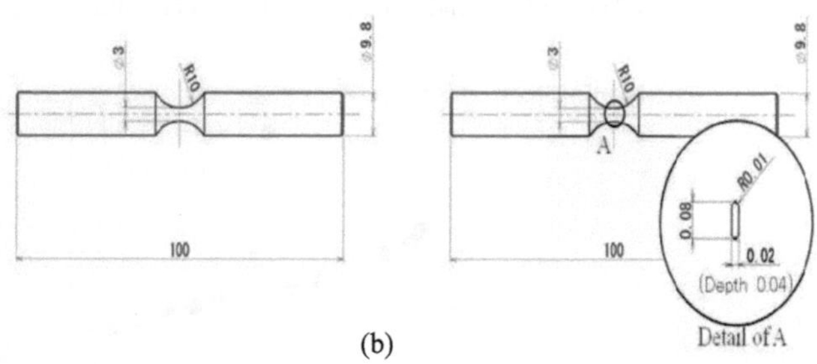

(a) (b)

Figure 2. Test specimens used in the present study.
(a) Plane specimen (Unnotched specimen)
(b) Notched specimen

2. Experimental procedures

Rotating bending fatigue tests were conducted in laboratory air at room temperature under a stress ratio of R = -1. Stress frequency employed was 5 - 15 Hz. To evaluate the fatigue lifetime, the plane specimen shown in Figure 2a was used. The fracture surfaces were observed using a scanning electron microscope (SEM) after the test. To evaluate the rate of fatigue crack growth of a short surface crack, rotating bending fatigue tests were conducted at the same condition as the plane specimen using the notched specimen shown in Figure 2b. The fatigue tests were interrupted at constant intervals to obtain replicas of the specimen surface. The crack lengths recorded on the replicas were measured using an optical microscope at magnification of 200 - 500. To calculate the maximum stress intensity factor Kmax, the following expression was used:

$$K_{max} = Y\ \sigma_a\ (\Pi\ a)^{0.5} \qquad (1)$$

where σa is the stress amplitude, a is a half crack length, and Y is a crack shape correction factor. A value of 0.73 was used for Y assuming that the crack shape is semicircular [6]. In addition to the observation of the replicas, the specimen surfaces were directly observed using SEM for investigation of crack growth path.

3. Results

3.1 S-N Curves

Figure 3 shows the experimental S-N data for two grades of fine grained WC-Co cemented carbides with different Co content at stress ratio of R = -1. The data with arrows in this figure indicates that the specimen was not broken until the cycles. From this figure, the S-N curves of both materials were similar, and the fatigue limits at R = -1 were estimated to be 1,500 MPa. However more in detail, the fatigue lifetimes for the WC-13%Co are little bit longer than those of the WC-18%Co under the constant stress amplitude.

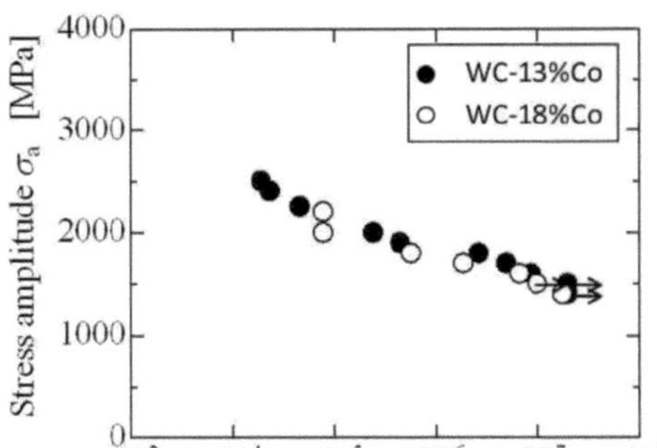

Figure 3. S-N curves for the fine-grained WC-Co cemented carbides.

3.2 Fractography

Figure 4 shows the SEM photographs [7] of the fracture surfaces for the WC-13%Co specimens failed at the stress amplitudes of 1,600 MPa (Figure 4a) and 1,700 MPa (Figure 4b), respectively. The right side of these photographs shows the enlarged one. In Figure 4a, as indicated by the white arrows in the figures, the crack initiates from the aggregate of the several WC grains with a size of 7 μm beneath the specimen surface. On the other hand, in Figure 4b, the fatigue crack initiates from the coarse WC grain with a size of 6 μm or so. So, the aggregate of several WC grains or coarse WC grain with sizes of 6 - 7 μm in diameter become the fatigue crack initiation site for this fine grained WC-Co material. The similar results were also observed for the WC-18%Co specimens.

(a) **σa = 1,600 MPa**

(b) **σa = 1,700 MPa**

Figure 4. Fatigue fracture surfaces of the WC-13%Co specimens showing the fatigue crack initiation sites. The right side is the enlarged photograph of the left side one [7].

Figure 5 shows the SEM photographs of the fracture surfaces of the specimens failed at the σa of 1,600 MPa (WC-13%Co) and at the σa of 1,700 MPa (WC-18%Co). The fracture surfaces near the fracture origin and the one in the center of the specimen are compared in these figures. The following difference between the former and the latter was noticeable in both materials; i.e., the fracture surface topography near the fracture origin was rough with some lacked portion of the WC grains, while the fracture surface topography in the center was relatively flat with many cleavage fractures of the WC grains.

(a) σ_a=1,600 MPa N_f = 8.8×10^6 cycles (WC-13%Co)

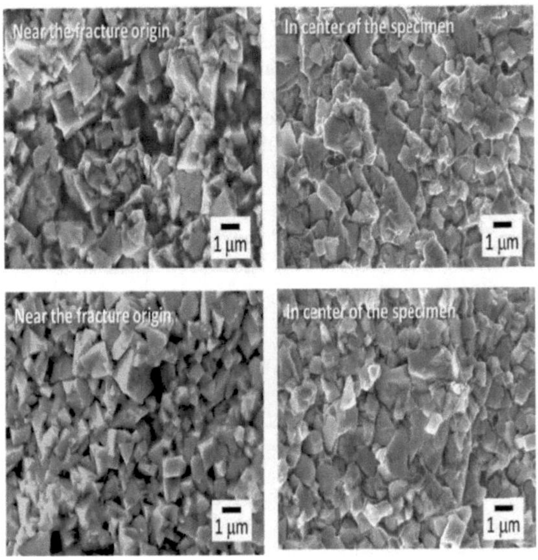

(b) **σ_a=1,700 MPa N_f = 1.9×10^6 cycles (WC-18%Co)**

Figure 5. Fatigue fracture surfaces of the specimens showing the comparison between near the fracture origin and in the center of the specimen.

3.3 Crack Growth behavior

Figure 6 shows the increase of crack lengths, 2*a* as a function of number of cycles, *N* for the both materials, WC-13%Co and WC-18%Co. The tests were conducted using the notched specimens under almost the same applied stress and initial crack length. The arrows in the figure show the data obtained just before final fracture of the specimens. As can be seen from the figure, the loading cycles from the crack initiation to the final fracture of the specimen are required more in the WC-18%Co than in the WC-13%Co. Further, in the former, the critical crack length prior to the specimen failure is longer than in the latter. These experimental data indicate that the crack growth resistance for the WC-18%Co is higher than that for the WC-13%Co within the range of 100 - 400 □m in crack length.

Figure 7 shows the relationship between the rate of crack growth d*a*/d*N* and the maximum stress intensity factor *K*max for the short surface fatigue crack. As can be seen from the figure, at the high *K*max region of 7 - 10 MPam$^{1/2}$, the crack growth resistance for the WC-18%Co is higher than that for the WC-13%Co. In the lower *K*max region below 6 MPam$^{1/2}$, the reverse trend, i.e., the crack growth resistance for the WC-13%Co is higher or equal to that for the WC-18%Co, is expected, but this point remain obscure due to the lack of the experimental data, further experiments are necessary.

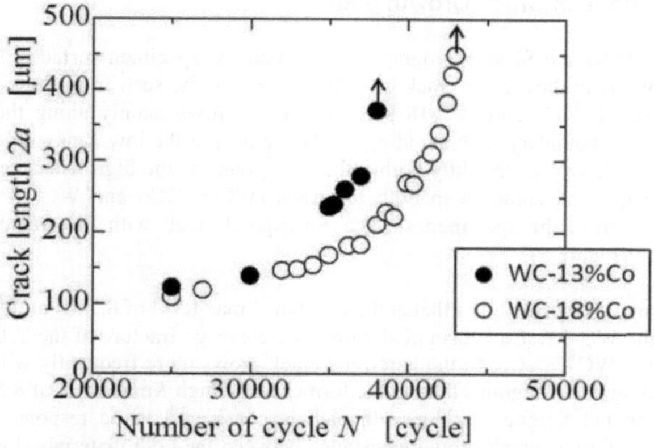

Figure 6. Crack growth behavior tested at σa=600 MPa (WC-13%Co) and σ_a=620 MPa (WC-18%Co).

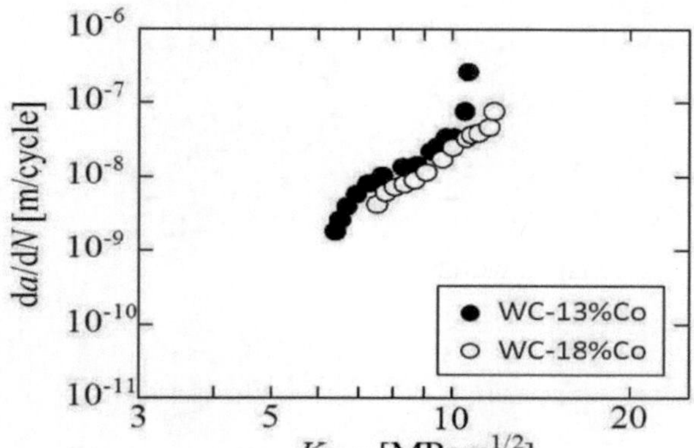

Figure 7. Relationship between d*a*/d*N* and *K*max.

3.4 Observations of Crack Growth path

Figure 8 shows the SEM photographs of the notched specimen surfaces observed at several values of *K*max during the crack growth test. As can be seen from these figures, in both alloys of WC-13%Co and WC-18%Co, the crack grows mainly along the WC/WC boundary or WC/Co boundary with avoiding the WC grains at the low *K*max region, on the other hand, the crack grows straightly within the WC grains at the high *K*max region. This finding was found to be common in both materials (WC-13%Co and WC-18%Co). The observation result from the specimen surface corresponds well with the fracture surface observation result (Figure 5).

Furthermore, it's worth noting that at the constant *K*max level of 8.5 MPam$^{1/2}$, the crack growth path in the WC-13%Co is occupied with more cleavage fracture of the WC grains as compared with the WC-18%Co. In the latter, the crack grows more frequently within the Co phase with avoiding the WC grains than in the former at the high *K*max level of 8.5 MPam$^{1/2}$. This difference in the fatigue crack growth path is considered to be responsible for the dissimilarity in the fatigue crack growth resistance between the both materials (Figure 6 and Figure 7).

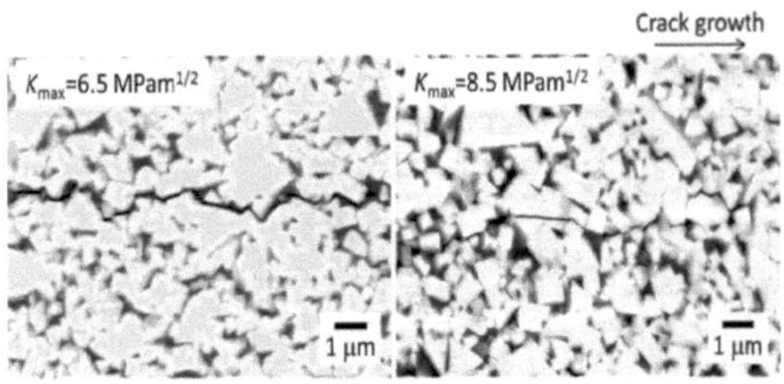

(a) WC-13%Co

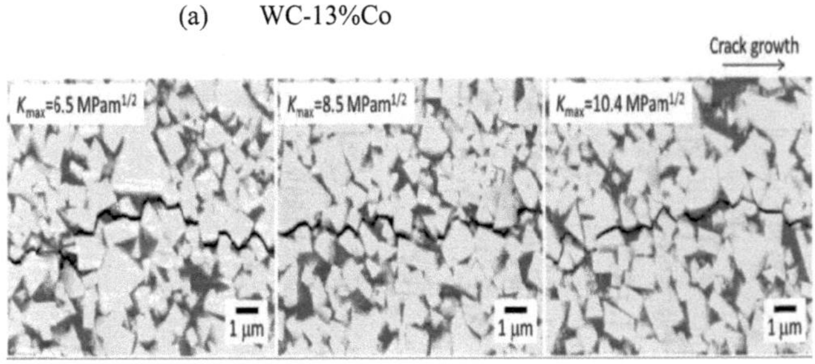

(b) WC-18%Co

Figure 8. Crack path appeared on the specimen surfaces at several *K*max.

4. Discussion

4.1 Effect of Co content on the crack growth resistance

From the fatigue crack growth path observations shown in Figure 8, the values of Kmax under which the cleavage fractures of the WC grains occur are estimated to be 8.5 MPam$^{1/2}$ and 10.4 MPam$^{1/2}$ for the WC-13%Co and WC-18%Co, respectively. Under these Kmax values, the rates of fatigue crack growth are almost the same, i.e., 2.5×10^{-8} m/cycle and 2.8×10^{-8} m/cycle for the WC-13%Co and WC-18%Co, respectively. This result indicates that the rate of fatigue crack growth is closely related with the crack growth path. In the material that has a higher Co content (WC-18%Co), crack driving force ahead of the crack tip is considered to be shielded by the plastic deformation in the binder phase of Co [8]. So, in the WC-18%Co, the crack does not have enough driving force to cleave the WC grain ahead of the crack tip until the value of Kmax reach the higher critical value (10.4 MPam$^{1/2}$) as compared with that (8.5 MPam$^{1/2}$) for the WC-13%Co.

4.2 S-N curves

Within the Co content, 13 –18 wt% used in the present study, the effect of the Co content on the fatigue lifetimes was minimal as shown in *S-N* curves of Figure 3. It was clarified that most of the fatigue lifetimes for the present hourglass-shaped specimen with a diameter of 3 mm is occupied with the short fatigue crack growth period from 5 - 10 μm [7]. During the short fatigue crack growth period (at the low Kmax region ($\approx$5 MPam$^{1/2}$)), the crack grows mainly along the WC/WC boundary or WC/Co boundary with avoiding the WC grains as shown in Figure 8. Therefor, under this situation, it is considered that the crack growth behavior is hardly affected by the Co phase. As a result, the similar *S-N* curves for the both materials were brought (Figure 3). In future, in order to improve the fatigue growth resistance and fatigue lifetimes, it is interesting to study whether further increase in the Co content is effective or not.

5. Conclusions

(1) S-*N* curves of both materials, WC-13%Co and the WC-18%Co were similar, and the fatigue limits at R = -1 were estimated to be 1,500 MPa. However more in detail, the fatigue lifetimes for the WC-13%Co are little bit longer than those of the WC-18%Co

(2) The aggregate of several WC grains or the coarse WC grain with sizes of 6 - 7 μm in diameter become the fatigue crack initiation sites for the both materials studied.

(3) The fracture surface topography near the fracture origin was rough with some lacked portion of the WC grains, while the fracture surface topography in the center was relatively flat with many cleavage fractures of the WC grains.

(4) At the high Kmax region of 7 - 10 MPam$^{1/2}$, the crack growth resistance for the WC-18%Co is higher than that for the WC-13%Co. In the lower Kmax region below 6 MPam$^{1/2}$, the reverse trend, i.e., the crack growth resistance for the WC-13%Co is higher or equal to that for the WC-18%Co, is expected, but this point remain obscure due to the lack of the experimental data.

(5) At the low *K*max region, the crack grows mainly along the WC/WC boundary or WC/Co boundary with avoiding the WC grains, on the other hand, at the high *K*max region, the crack grows straightly within WC grains. This finding was found to be common in both materials.

(6) In the WC-18%Co, the crack grows more frequently within the Co phase with avoiding the WC grains than in the WC-13%Co at the high *K*max level of 8.5 $MPam^{1/2}$.

References

I. Konyashin, "Cemented Carbides for mining, Comprehensive Constru **Hard Materials**, 1(2014), 425-451.

2. A.H. Li, J. Zhao, D. Wang, X.L. Gao, and H.W. Tang, "Three-point bending fatigue behavior of WC-Co cemented carbides,"**Materials & Design**, 45 (2013) 271-278.

3. T. Sailer, M. Herr, H.G. Sockel, R. Schulte, H. Feld, and L.J. Prakash, "Micro mechanical properties of ultrafine-grained hardmetals,"**International Journal of RefractoryMetals & Hard Materials**, 19 (2001) 553-559.

4. L. Llanes, Y. Torres, and M. Anglada, "On the fatigue-Co crac cemented carbides: kinetics description, **Acta mic Materialia**, 50 (2002) 2381-2393.

5. M.H. Boo, and C.Y. Park, "Mechanisms of-Co cemented fatigue carbides,"Advances in Fracture and Strength, Pts 1- 4, 297-300 (2005) 1120-1125.

6. Y. Murakami, and M. Ishida, Trans **Jpn Soc Mech Eng** A, 51(1987) 1050-1056.

7. H. Mikado, S. Ishihara, N. Oguma, K. Masuda, and S. Kawamura, growth behavior of WC-Co Cemented Advanced carbide," **Materials Research**, 891-892(2014), 955-960.

8. L.S. Sigl and H.F. Fischmeister, **Acta Metallurgica,** 36 (1988) 887-897.

Fatigue of Materials III: Advances and Emergences in Understanding
Edited by: T.S. Srivatsan, M. Ashraf Imam, and R. Srinivasan
TMS (The Minerals, Metals & Materials Society), 2014

EXPERIMENTAL AND NUMERICAL STUDIES ON MULTISTAGE STRENGTH DEGRADATION IN NOTCHED CONCRETE BEAMS UNDER REPEATED LOADS: A REVIEW

Masaaki Nakano[1], Zihai Shi[*1], Yukari Nakamura[1], Cuiping Liu[1], Hiroshi Tanaka[1]

[1]Research and Development Centre
Nippon Koei Co., Ltd.
2304 Inarihara, Tsukuba
Ibaraki 300-1259, **Japan**
Email*: a4739@n-koei.co.jp (SHI)

Abstract

The multistage strength degradation theory states that fatigue is caused by the sporadic sudden change of cracking behaviour in a system under cyclic loading, leading to intermittent strength reduction of the system and its eventual failure. To prove this theory, an experimental study on the load-carrying capacity of notched concrete beams was carried out first, in which the sizes of several notches were enlarged sequentially to simulate the growth of cracks under repeated loading. A multistage strength degradation relation between the maximum load and notch size was obtained. Next, numerical studies of the same problem using sequential loads for propagating multiple cracks in the beam were performed, and the results presented several multistage strength degradation processes based on the numerical simulation of crack propagation. This review paper introduces these experimental and numerical studies on this newly-established fatigue theory, with an emphasis on the latter.

1. Introduction

In a numerical study of the load-carrying capacity of notched concrete beams, which required crack analyses of multiple cracking behaviours, the maximum load was found to strongly depend on the failure mode [1]. The loading conditions and notch arrangements of the simple beams in the original study are illustrated in **Figure 1**. As shown, the beam contained three notches, A, B and C, and a single load was applied to the side-notch A in **Figure 1(a)** and the side-notch C in **Figure 1(b)**, respectively. While notches A and B were kept at a constant size of 10 mm, notch C was enlarged from 10 mm to 150 mm to study the relations between the maximum loads and failure modes under the given eccentric load conditions. The material properties for the numerical study are shown in **Table I**: elastic modulus E, Poisson's ratio ν, compressive strength f_c, tensile strength f_t, and fracture energy G_F.

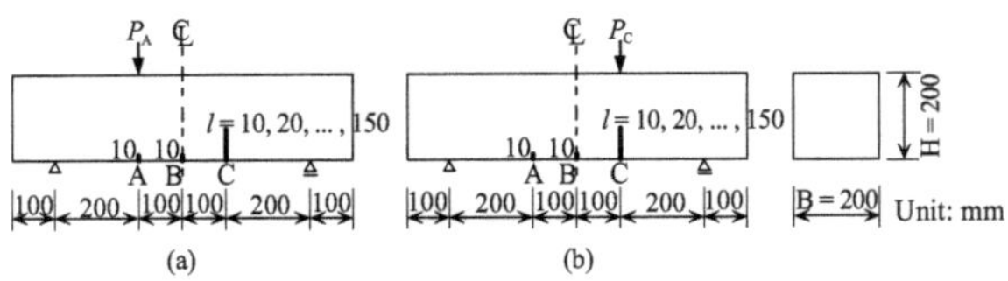

Figure 1. Numerical models of fracture tests on notched concrete beams under eccentric loads: (a) a single load applied at side-notch A; (b) a single load applied at side-notch C.

Table I. Material properties of concrete used in numerical studies.

E (kN/mm²)	ν	f_c (kN/mm²)	f_t (kN/mm²)	G_F (N/mm)
26.6	0.2	37.5	2.86	0.1

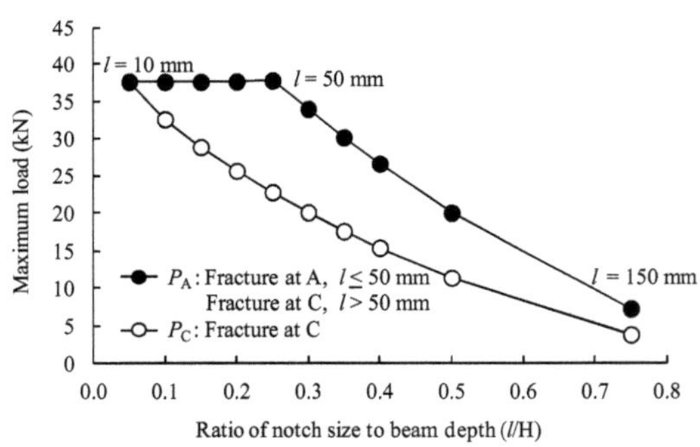

Figure 2. Relationship between maximum load and the ratio of notch size to beam depth.

The obtained relations are shown in **Figure 2**, which contains two curves. When the eccentric load was applied at notch C, a monotonically decreasing relation between the maximum load and the size of notch C was obtained, and the dominating crack for beam failure was shown to invariably propagate from notch C. On the other hand, when the eccentric load was applied at notch A, the obtained maximum load was unaffected by the enlargement of notch C until it reached a critical value, beyond which the peak load diminished quickly with a reduction rate faster than that of the P_C curve. It was shown that two failure modes were involved in obtaining the maximum load P_A. Before exceeding the threshold value of notch C at 50 mm, the dominating crack for the beam failure originated from notch A; beyond that point it propagated from notch C. As shown by this study, a change of cracking behaviour and failure mode may significantly reduce the load-carrying capacity of a simple beam.

In studies of metal fatigue, it has long been known that the fatigue strength of a test specimen is not affected by introducing artificial micro holes into the specimen unless the size of these artificial defects exceeds a critical value, beyond which there is a significant reduction in fatigue strength [2]. **Figure 3** illustrates two types of carbon steel specimen and the shape of twelve identical micro holes, which were introduced into these specimens at three locations (A, B and C) with an equal spacing of 5 mm: at each location, along the perimeter of a cylindrical specimen, four micro holes were drilled into the specimen at equal intervals. Details on the material properties of the test specimens and test procedures can be found elsewhere [2].

Relations between the experimentally obtained fatigue limit (the stress level below which a test specimen will never break no matter how many stress cycles are applied) and the size of micro holes under rotating bending are shown in **Figure 4**. As seen, a critical diameter for these micro holes exists for each type of material, which is approximately 70 μm for S10C steel and 35 μm for S45C steel. By introducing micro holes that are smaller than the critical size into a specimen, no particular influence on fatigue strength can be observed and the fatigue limit for smooth specimens (without micro holes) is obtained; otherwise, a reduction in the fatigue limit occurs. It is known that the critical sizes of 70 μm and 35 μm are closely related to the fact that the maximum non-propagating crack observed at the fatigue limit in annealed, smooth specimens is approximately 100 μm for S10C steel and 50 μm for

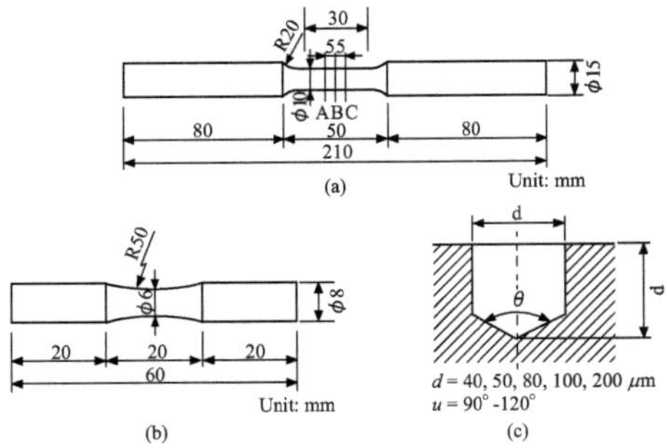

Figure 3. **Test specimens and the shape of micro holes from fatigue tests [2].**

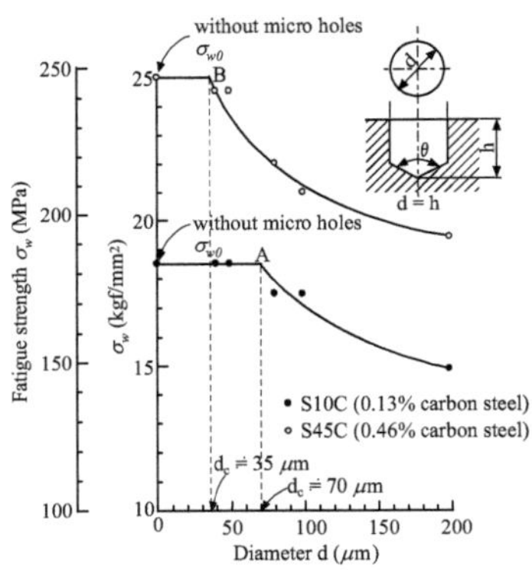

Figure 4. **Relations between fatigue limit and size of micro holes [2] (courtesy of Yokendo).**

S45C steel under rotating bending. Obviously, fatigue strength degradation beyond the threshold region indicates fundamental changes in the fracture behaviour of test specimens having larger micro holes, as is inferred from the previous study on the strength degradation of notched concrete beams.

In general, initial flaws and defects exist in all engineering materials; only the degree of imperfection varies. Under cyclic loading, the material weakening process in a structural or mechanical system inevitably involves multiple cracking originated from some of these spatially distributed initial imperfections, and diverse cracking behaviours are expected [2-9]. As a threshold value in terms of a critical crack length or a critical crack density (such as the maximum number of cracks in a certain location), etc. is reached, new cracking behaviours may abruptly emerge and replace the previous ones, causing strength degradation. As fatigue is a slow but continuous fracture process, it is logical to assume that this unique fracture mechanism of strength degradation repeatedly occurs under cyclic loading, thus forming a multistage strength degradation process leading to fatigue failure.

To prove this new theory, an experimental study on the load-carrying capacity of notched concrete beams was carried out first, in which the sizes of several notches in the test specimens were enlarged sequentially to simulate the growth of cracks under repeated loads. A multistage strength degradation relation between the maximum load and notch size was obtained [10]. Next, numerical studies of the same problem using sequential loads for propagating multiple cracks in the beam were performed, and the results revealed several multistage strength degradation processes based on the numerical simulation of crack propagation [11]. In this review paper, these experimental and numerical studies on this newly-established fatigue theory are introduced, with an emphasis on the numerical analysis of notched concrete beams under sequential loading.

2. Fracture Tests of Notched Concrete Beams with Increasing Notch Sizes

Figure 5 shows a notched concrete beam under four-point bending; fracture tests were carried out to investigate how increasing the notch size affected the load-carrying capacity of the beam. The mix proportions of the concrete are summarised in **Table II**, and its material properties are presented in **Table III**. As shown in **Figure 5**, the beam contained five initial notches equally spaced at an interval of 100 mm, except the spacing between notches A and E, which was deliberately set at a smaller value of 80 mm to embed an unsymmetrical effect in the problem. The test specimens were divided into three groups, and **Table IV** lists the notch size arrangement in each group. Initially, all the notches were given a uniform size of 10 mm. Then, a single notch was selected from each group to represent a propagating crack with increasing crack lengths. In group 1 notch D served as the pseudo-crack with three sizes of 30, 50 and 60 mm. In group 2, while keeping notch D at 60 mm, notch E assumed two sizes of 40 and 60 mm. In group 3, while keeping notches D and E at 60 mm, notch B was enlarged from 20 mm to 40 mm. These notch size arrangements were determined based on preliminary numerical studies of the fracture tests. Note that for each test case with a specific notch size arrangement, three identical specimens were prepared for the tests.

Figure 6 shows the experimental results on the relationship between the maximum load and notch size, as well as the fracture modes of the test specimens. The horizontal axis represents the continuing enlargement of notch size in the order of notches D, E and B. Despite the scatter in the test

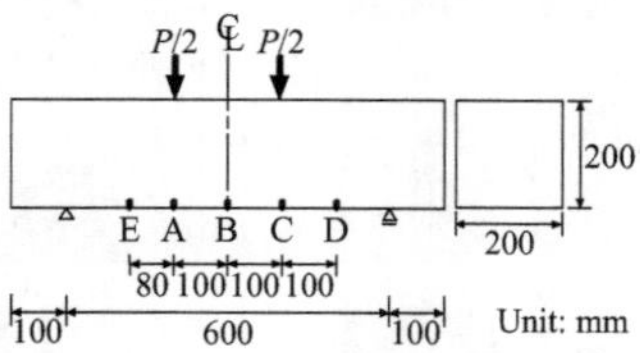

Figure 5. **Fracture tests on the load-carrying capacity of notched concrete beams under four-point bending by increasing notch sizes.**

Table II. Composition of concrete used in fracture tests.

Maximum size of coarse aggregate (mm)	Slump (cm)	Air (%)	W/C ratio (%)	Sand-coarse aggregate ratio (%)
15	11.5	4.8	55	46

Water W (kg/m^3)	Cement C (kg/m^3)	Fine aggregate S (kg/m^3)	Coarse aggregate G (kg/m^3)	Admixture AE (g/m^3)
178	324	810	965	16.20

Table III. Material properties of concrete used in fracture tests.

E (kN/mm^2)	v	f_c (kN/mm^2)	f_t (kN/mm^2)	G_F (N/mm)
29.2	0.18	34.3	3.17	0.142

Table IV. Notch size (dimensions in mm) arrangements in fracture tests of notched concrete beams.

Group \ Notch	E	A	B	C	D
1	10	10	10	10	30, 50, 60
2	40, 60	10	10	10	60
3	60	10	20, 40	10	60

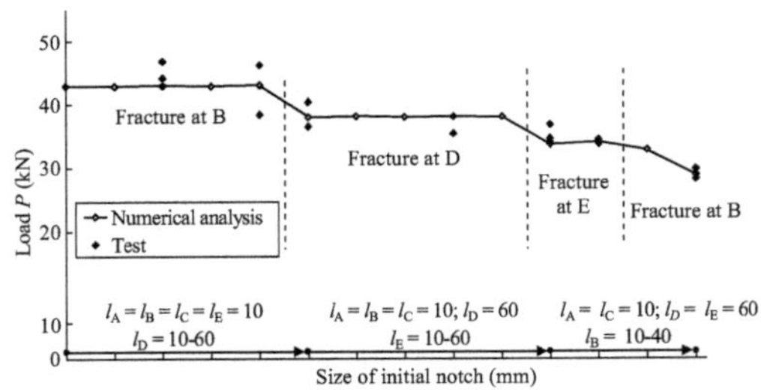

Figure 6. Experimental and numerical results on multistage strength degradation due to change of fracture behaviour in bending tests of notched concrete beams.

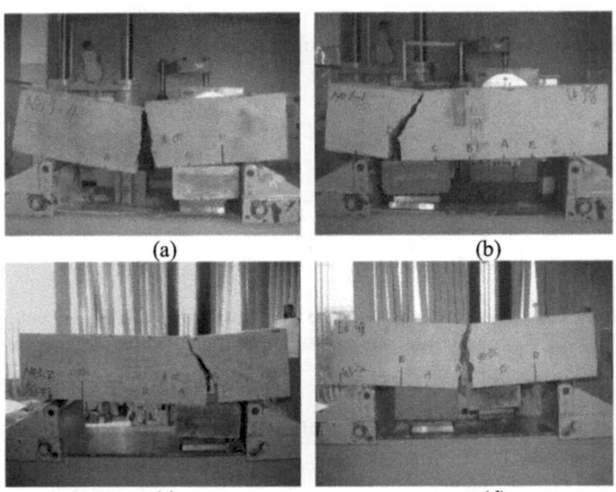

(a) (b) (c) (d)

Figure 7. Photos of fractured test specimens: (a) fractured at notch B; (b) fractured at notch D; (c) fractured at notch E; (d) fractured at notch B.

results, the general trend clearly shows the reduction of beam strength when notch D reached 60 mm, notch E reached 60 mm, and notch B reached 40 mm, accompanied by the transitions of failure modes shown in **Figure 7**. As a result, the load-carrying capacity of the beam decreased through multiple stages, which was first reported in [10]. Numerical analyses were also carried out to reproduce the fracture processes and calculate the maximum loads, which both compared well with the test results. According to the numerical analyses, the third stage of strength degradation should occur when notch B reaches 30 mm.

The importance of this study is twofold. Firstly, it reconfirms the previous findings that, for a given load condition, increasing the size of an initial notch may not affect the fracture behaviour and maximum load, provided it is less than a threshold value. When it exceeds that critical size, however, a fundamental change in fracture behaviour can take place, and the load-carrying capacity can drop significantly. Secondly, the study shows that this strength degradation mechanism can repeatedly occur, i.e., the load-carrying capacity can decrease in multiple stages, and at each stage a unique cracking behaviour governs the fracturing processes while the strength of the previously weakened material remains basically unchanged. Even though these observations are based on studies of notched beams, their implications for understanding the mechanism of strength degradation during cyclic loading are obvious.

3. Numerical Studies of Notched Concrete Beams under Sequential Loading

3.1 Scheme of Numerical Study

As a further investigation on the theory of multistage strength degradation, **Figure 8** shows the scheme of a numerical study by employing the same notched beam of **Figure 5**, but an important difference from the previous study is that crack propagation is simulated by loading the beam, not by enlarging the notch size. For notches A, B and C the notch size is 10 mm, and for notches D and E it is 20 mm. To propagate cracks, sequential loads are applied above the initial notches in four loading steps, following a random order of D, E, C and A, as shown in **Figure 8(b)**. In each loading step, a single load is applied to extend the crack from the notch below to a certain length before moving to the next loading point. Obviously, during each of the four loading processes multiple cracks may propagate from other notches, too. At each step of crack growth the load-carrying capacity of the beam under three-point bending is calculated through crack analysis, as schematically illustrated in **Figure 8(c)** where the cracks at the initial notches are treated as pre-existing cracks for strength evaluation. Hence, crack analysis is carried out alternately using two loading systems: sequential loading and three-point bending. The FE model of the notched beam is shown in **Figure 9**, and a

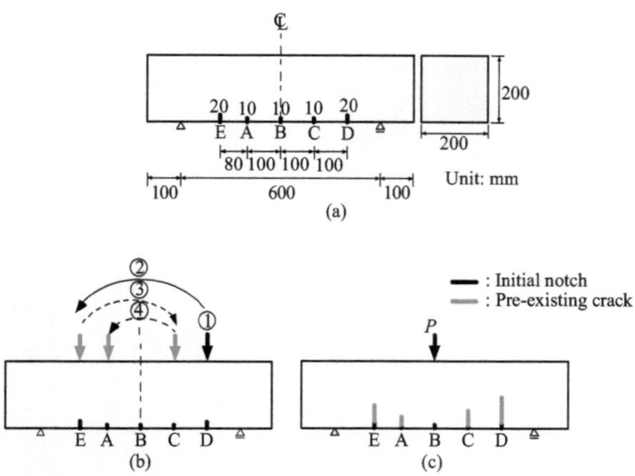

Figure 8. **Scheme of numerical studies on a notched beam: (a) beam geometry; (b) sequence of loading; (c) computation of load-carrying capacity of the notched beam with pre-existing cracks.**

bilinear tension-softening relation for crack analysis is illustrated in **Figure 10**. The material properties of **Table I** are used for this numerical study.

To facilitate the convergence of numerical solutions amid the load point movements, the crack path from an initial notch is fixed and is assumed to be vertical. This assumption precludes the need for crack-path remeshing and makes the task of setting the pre-existing cracks at the initial notches under three-point bending much easier. However, this simplification may also cause some minor inaccuracy in the crack analysis, which will be addressed when discussing the results of numerical analyses.

The extended fictitious crack model (EFCM) is employed for crack analysis, which solves a multi-crack problem discretely through crack equations that are formulated based on relevant cracking modes; for details of the EFCM and its numerical implementation, refer to the previous publications [10, 12]. Also, the singularity issue encountered when formulating the EFCM is clarified in [13]. The open-source computer program, CAIC-M1.FOR [14], is used for the crack analysis in the following numerical studies.

4. Results and Discussion

Figures 11 to **14** present four selected scenes of crack propagation as examples to summarise the results of crack analysis under sequential loading and three-point bending. Note that these cases are connected to two threshold regions where the transition of failure mode and the decrease of load-carrying capacity under three-point bending take place, as the size of a propagating crack approaches

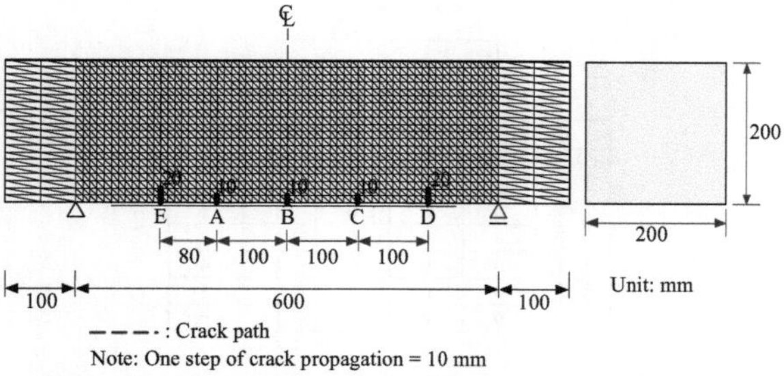

Figure 9. FE model of the notched concrete beam used for crack analysis.

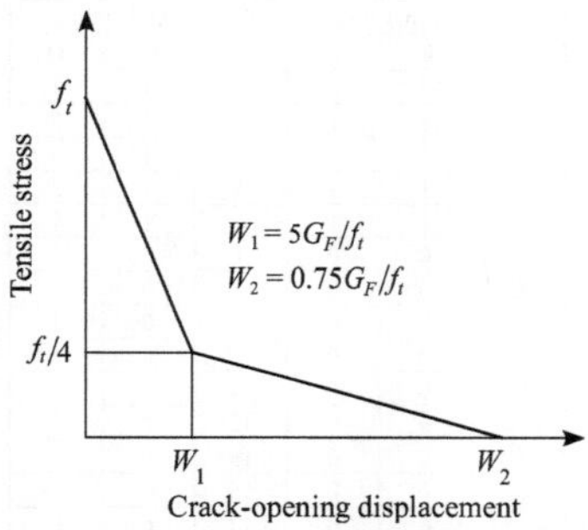

Figure 10. Bilinear tension-softening relation of concrete.

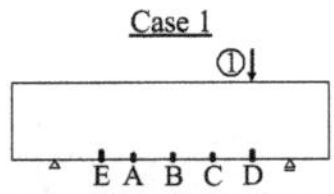

Crack propagation under sequential loading					
Crack	E	A	B	C	D
Step of crack propagation	0	0	0	0	1
	0	0	0	1	2
	0	0	0	1	3
	0	0	0	1	4
	0	0	0	1	5
	0	0	0	1	6

(a)

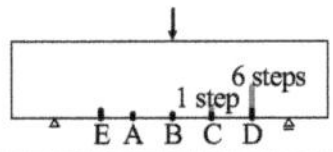

Crack propagation under three-point bending					
Crack	E	A	B	C	D
Step of crack propagation	0	0	1	1	1
	0	1	2	1	2
	1	1	4	1	3
	1	1	5	1	3
	1	1	6	1	2
	0	0	7	0	2
	0	0	8	0	2
	0	0	9	0	1
	0	0	10	0	1
	0	0	11	0	1
	0	0	12	0	1
	0	0	13	0	1
	0	0	16	0	0
	0	0	17	0	0
	0	0	18	0	0

Note: Beam failure at B; Jump at steps 3, 14, 15

(b)

Figure 11. Results of crack analysis in case 1: (a) crack propagation due to sequential loading; (b) computation of load-carrying capacity under three-point bending with pre-existing cracks.

a critical value. The obtained relation between the load-carrying capacity and crack length is shown in **Figure 15**, which clearly exhibits multistage strength degradation.

In case 1 of **Figure 11(a)**, with a single load applied above notch D a crack propagates from that notch for six steps, accompanied by a one-step growth of a crack from notch C. No cracks emerge from other notches. Based on the FE mesh shown in **Figure 9**, it is known that one step of crack growth equals the spacing between two neighbouring nodes, i.e., 10 mm. As shown in **Figure 11(b)**, the initial conditions set for calculating the load-carrying capacity under three-point bending are illustrated in the notched beam by two pre-existing cracks, i.e., a crack of 10 mm at notch C (one step of propagation) and a crack of 60 mm at notch D (six steps of propagation). To account for the cumulative damage effect by cyclic loads on the crack surfaces, i.e., smoothing out the rough crack surfaces due to their repeated opening and closing, the cohesive forces at the pre-existing cracks are deliberately ignored in calculating the beam strength. The results of crack analysis show that the potential failure mode is beam fracture at notch B, and there is no strength degradation of the beam up to this stage of crack propagation, as shown in **Figure 15**.

Case 2

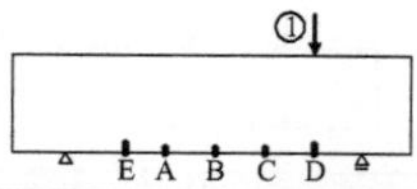

Crack propagation under sequential loading					
Crack	E	A	B	C	D
Step of crack propagation	0	0	0	0	1
	0	0	0	1	2
	0	0	0	1	3
	0	0	0	1	4
	0	0	0	1	5
	0	0	0	1	6
	0	0	0	0	7

(a)

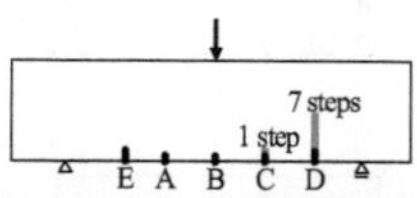

Crack propagation under three-point bending					
Crack	E	A	B	C	D
Step of crack propagation	0	0	1	0	1
	0	1	3	1	3
	0	1	3	1	4
	0	1	2	1	5
	0	0	2	0	6
	0	0	1	0	7
	0	0	1	0	8
	0	0	0	0	9
	0	0	0	0	10

Note: Beam failure at D; Jump at step 2

(b)

Figure 12. Results of crack analysis in case 2: (a) crack propagation due to sequential loading; (b) computation of load-carrying capacity under three-point bending with pre-existing cracks.

In case 2 of **Figure 12(a)**, if crack D propagates one step further than in case 1 as the load at notch D is increased, the failure mode of the notched beam will abruptly change and the beam will break up at notch D, as shown in **Figure 12(b)**. Based on the strength-degradation graph of **Figure 15**, this failure mode transition will decrease the load-carrying capacity by approximately 5% from the previous strength level.

Case 3 of **Figure 13** and case 4 of **Figure 14** display cracking behaviours of the beam in the second stage of the sequential loading, i.e., while retaining the previous crack history, the sequential load is now moved to notch E to propagate cracks from that side of the beam. As a result, two new cracks emerge at notches E and A, while the older cracks at notches D and C are forced to close due to local compression at these cracks. As with cases 1 and 2, cases 3 and 4 are also two consecutive cases in terms of crack growth: crack E propagates five steps in case 3 (**Figure 13(a)**) and six steps in case 4 (**Figure 14(a)**). In both cases crack A propagates only one step.

Since material damage due to fracture is irreversible, it is clear that in calculating the load-carrying capacity of the beam the older cracks from the previous loading history should also be

Case 3

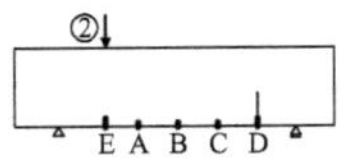

Crack propagation under sequential loading					
Crack	E	A	B	C	D
Step of crack propagation	0	0	0	0	1
	0	0	0	1	2
	0	0	0	1	3
	0	0	0	1	4
	0	0	0	1	5
	0	0	0	1	6
	0	0	0	0	7
	1	0	0	0	0
	2	1	0	0	0
	3	1	0	0	0
	4	1	0	0	0
	5	1	0	0	0

(a)

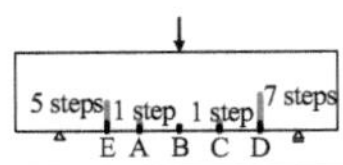

Crack propagation under three-point bending					
Crack	E	A	B	C	D
Step of crack propagation	1	0	1	0	1
	3	1	3	1	3
	3	1	3	1	4
	3	1	2	1	5
	2	0	1	0	6
	2	0	1	0	7
	1	0	0	0	8
	1	0	0	0	9
	0	0	0	0	10

Note: Beam failure at D; Jump at step 2

(b)

Figure 13. **Results of crack analysis in case 3: (a) crack propagation due to sequential loading; (b) computation of load-carrying capacity under three-point bending with pre-existing cracks.**

treated as pre-existing cracks as with the new cracks (even though they are closed under the new load system). Therefore, the notched beam under three-point bending contains four pre-existing cracks in case 3 and case 4, as shown in **Figure 13(b)** and **Figure 14(b)**. The sizes of the pre-existing cracks in the order of E, A, C, D are 50 mm, 10 mm, 10 mm and 70 mm for case 3, and 60 mm, 10 mm, 10 mm and 70 mm for case 4. As the results of crack analysis under three-point bending show, the beam experiences a transition in failure mode as the sequential load extends crack E from five steps in case 3 to six steps in case 4, i.e., from a notch D fracture to a notch E failure. As shown in **Figure 15**, the load-carrying capacity remains constant up to the fifth step of propagation by crack E, and then drops suddenly by approximately 5% at the crack's sixth step of extension.

As the sequential loading continues at notch C and then at notch A to propagate more cracks at different locations, the load-carrying capacity of the beam is shown to decrease in a similar fashion. For details of the numerical analyses, refer to **Figure 15**. Obviously, the obtained multistage strength degradations can continue as more loads are added to the sequential loading. In reality, however, this process of strength degradation under repeated loading cannot continue forever, because an abrupt

Case 4

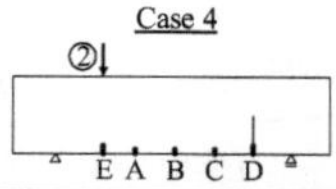

Crack propagation under sequential loading					
Crack	**E**	**A**	**B**	**C**	**D**
Step of crack propagation	0	0	0	0	1
	0	0	0	1	2
	0	0	0	1	3
	0	0	0	1	4
	0	0	0	1	5
	0	0	0	1	6
	0	0	0	0	7
	1	0	0	0	0
	2	1	0	0	0
	3	1	0	0	0
	4	1	0	0	0
	5	1	0	0	0
	6	0	0	0	0

(a)

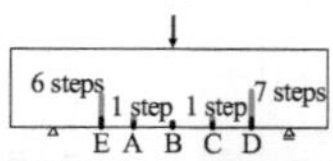

Crack propagation under three-point bending					
Crack	**E**	**A**	**B**	**C**	**D**
Step of crack propagation	1	0	1	0	1
	2	0	1	1	2
	3	0	2	1	3
	4	0	2	1	3
	5	0	2	1	2
	6	0	1	1	2
	7	0	1	0	2
	8	0	0	0	1
	9	0	0	0	1
	10	0	0	0	0
	11	0	0	0	0

Note: Beam failure at E

(b)

Figure 14. **Results of crack analysis in case 4: (a) crack propagation due to sequential loading; (b) computation of load-carrying capacity under three-point bending with pre-existing cracks.**

structural failure will terminate the process when the much-reduced load-carrying capacity of the beam can no longer sustain the design loads.

Concerning the failure modes in **Figure 15**, as crack C propagates for the first three steps under the sequential load, the failure mode under three-point bending changes, i.e., from E failure to B failure then to E failure again. This is believed to be the result of adopting a fixed-path strategy for crack analysis in the present study. If crack path remeshing is employed, the beam is expected to fail only at notch E at this stage of crack propagation. This is a minor inaccuracy, however, judging from the fact that the change of failure mode from E to B has little influence on the maximum load, as shown in **Figure 15**. In other words, these two failure modes are competing cracking modes, and with a slight change of condition, either of them can become the actual mode of failure. **Figure 16** presents the load-displacement relations obtained under three-point bending for the notched beam with pre-existing cracks. As expected, the greatest reduction in structural stiffness occurs at the moment the failure mode changes, as is clearly shown in each of the four figures that correspond to the four sequential loads.

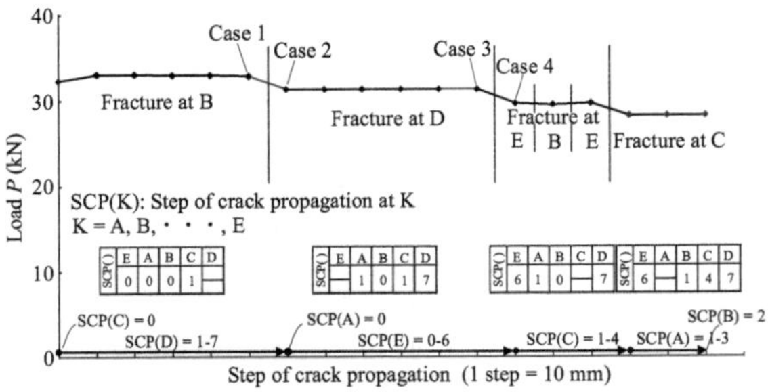

Figure 15. Numerical results on multistage strength degradation due to change of cracking behaviour and potential failure mode in a notched beam.

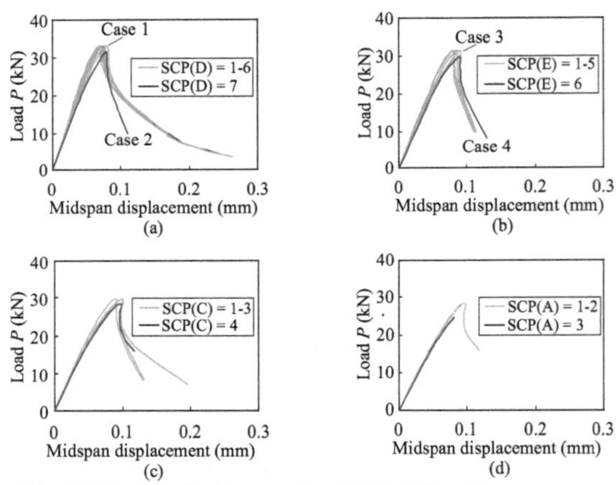

Note: SCP(K) = Step of crack propagation at K; K = A, B, ···, E.

Figure 16. Load-displacement relations of notched concrete beams with pre-existing cracks under three-point bending: (a) pre-existing cracks due to sequential load 1; (b) pre-existing cracks due to sequential load 2; (c) pre-existing cracks due to sequential load 3; (d) pre-existing cracks due to sequential load 4.

Though the strength degradation relation in **Figure 15** involves transitions of failure mode, strength degradation can also take place under a single failure mode, as demonstrated by the following two cases. The strength degradation curve in **Figure 17** is obtained by slightly enlarging notch B in **Figure 8(a)** to 20 mm and changing the sequential loads. Under these conditions, it is found that all the potential failure modes under three-point bending are the mid-span failure (fracture at B). Even though there is no change in the potential failure mode here, the occasionally spontaneous propagation of crack B amid various cracking activities in the beam does represent the sudden emergence of a new cracking behaviour, which causes strength degradation. In the next example of **Figure 18**, with a size arrangement of 20 mm for notch B and 10 mm for the rest, the central notch becomes the weakest point for stress concentration and crack propagation. However, under the random sequential loads applied at the surrounding notches of notch B, numerical analysis shows that strength degradation still

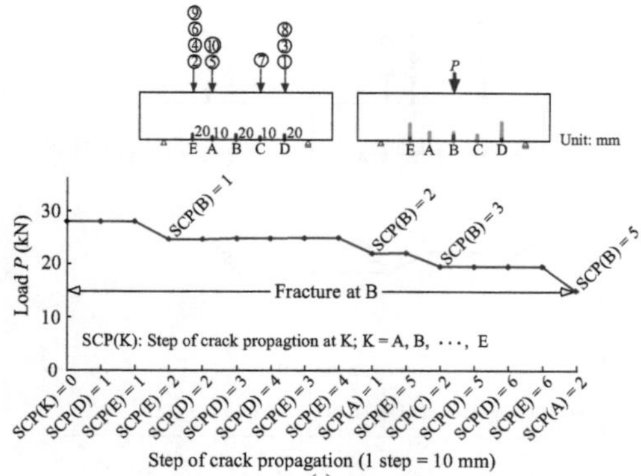

(a)

Crack propagation under sequential loading

SPC(K) / Loading step	E	A	B	C	D
①	0	0	0	0	1
②	1	0	0	0	1
	2	0	1	0	1
③	2	0	1	0	2
	2	0	1	1	3
	2	0	1	1	4
④	3	0	1	1	4
	4	0	1	1	4
⑤	4	1	2	1	4
⑥	5	1	2	1	4
⑦	5	1	3	2	4
⑧	5	1	3	2	5
	5	1	3	2	6
⑨	6	1	3	2	6
⑩	6	2	5	2	6

(b)

Figure 17. Example 1 on multistage strength degradation without change of potential failure mode: (a) multistage strength degradation; (b) crack propagation under sequential loading.

takes place through a multistage process that involves active crack propagation from the other notches, as well as the intermittent growth of crack B. Only one failure mode is involved in the process, i.e., fracture at B. These two cases may shed light on the fatigue failure of notched test specimens under cyclic loading, which usually fail at the notch section. Although this type of failure appears to be the work of a single crack, these two case studies reveal the actual fatigue process as involving multiple cracking activities and multistage strength degradations.

Concrete is a heterogeneous material consisting of aggregates and cement pastes bonded together at the interface, and the material is inherently weak in tension due to the limited bonding strength and various pre-existing micro cracks and flaws formed during hardening of the matrix. As such, fracture of concrete involves complicated micro-failure mechanisms that include micro cracking, crack deflection, crack branching, crack coalescence and debonding of the aggregate from the matrix. Obviously, each of these micro-failure mechanisms can extensively develop under repeated loads, and their transitions (from one form to another in hierarchical order of crack formation) at certain threshold points of crack formation and growth may signify the emergence of new cracking behaviours and new failure modes that cause strength degradation.

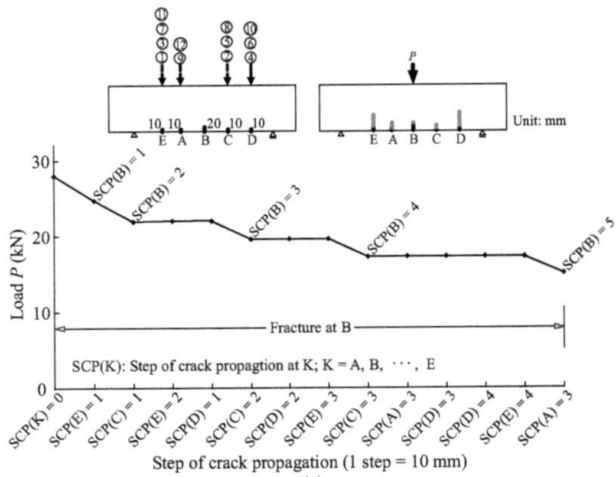

(a)

Crack propagation under sequential loading

Loading step \ SPC(K)	E	A	B	C	D
①	1	1	1	0	0
②	1	1	2	1	0
③	2	2	2	1	0
④	2	2	2	1	1
⑤	2	2	3	2	1
⑥	2	2	3	2	2
⑦	3	2	3	2	2
⑧	3	2	4	3	2
⑨	3	3	4	3	2
⑩	3	3	4	3	3
	3	3	4	3	4
⑪	4	3	4	3	4
⑫	4	3	5	3	4

(b)

Figure 18. Example 2 on multistage strength degradation without change of potential failure mode: (a) multistage strength degradation; (b) crack propagation under sequential loading.

It is known that metal fatigue is basically associated with cyclic slip in the lattice structure of metallic materials and its conversion into crack initiation and crack extension. Despite the fundamental differences between the micro deformation mechanisms of concrete and metals, cracks and their evolutionary growths under cyclic loading are the common cause of fatigue failure in both types of material. Hence, given the detailed fatigue process, the multistage strength degradation theory that focuses on the transition of fracturing behaviour and the degradation of strength during fatigue, can be used to explain the fundamental mechanism of fatigue not only for concrete, but also for metals. Regarding the numerous studies on the mechanism of fatigue, the following comment is quoted [15]: *According to Ekobori [16], existing theories and suggested fatigue mechanisms just describe the known experimental facts in analytical form. His conclusions are pessimistic in that he indicates that neither now nor in the near future might we expect a detailed and correct explanation of all aspects of fatigue using the same (e.g, atomistic or phenomenological) notions. Reasons for such pessimism persist because, despite very important results of experimental studies in this area, the majority of failures in practice are associated particularly with fatigue.* In the study of fatigue, the focus should be placed not only on the fatigue process that shows how cracks initiate and grow, but also on how the strength degradation takes place during that process. The multistage strength degradation theory focuses on both, and thus reveals the fundamental mechanism of fatigue.

5. Summary

1. In the material weakening process of concrete under fatigue, strength decreases in successive stages amid multiple cracking activities that involve changes in the cracking behaviours and with or without change in the potential failure mode.
2. Under cyclic loads a threshold value of crack propagation exists, and the load-carrying capacity of a structural member can remain constant until this value is reached. Beyond it, new cracking behaviours abruptly emerge and the load-carrying capacity is reduced.
3. This strength degradation can take place in multiple stages, and at each stage a very different cracking behaviour may dominate the fracturing process while the maximum load remains basically unchanged. This process of strength degradation can continue until the residual strength of the material can no longer sustain the level of stress produced by the design load, leading to an abrupt structural failure.
4. Despite the fundamental differences between the micro deformation mechanisms of concrete and metals, cracks and their evolutionary growths under cyclic loading are the common cause of fatigue failure in both types of material. Hence, given the detailed fatigue process, the multistage strength degradation theory can be used to explain the fundamental mechanism of fatigue for both materials.

References

1. Z. Shi, and M. Suzuki, **Int. J. Construction and Building Materials**, 18, 173-80 (2004).

2. Y. Murakami, Metal Fatigue: Effects of Small Defects and Nonmetallic Inclusions, p 33-53, Yokendo, Tokyo (1993; in Japanese).

3. S. To, S. B. Lambert, and D. J. Burns, **Int. J. Fatigue**, 15, 333-40 (1993).

4. A. Martin-Meizoso, J. M. Martinez-Esnaola, and M. Fuentes-Perez, **Int. J. Theoretical Applied Fracture Mechanics,** 23, 219-33 (1995).

5. S. Pitt, and R. Jones, **Int. J. Engineering Failure Analysis**, 4, 237-57 (1997).

6. Z. Wu, and X. Sun, **Int. J. Engineering Fracture Mechanics**, 59, 253-59 (1998).

7. E. B. Shields, **Int. J. Fatigue**, 23, 13-20 (2001).

8. C. Wessel, A. Cisilino, O. Santi, J. Otegui, and M. Chapetti, **Int. J. Theoretical Applied Fracture Mechanics**, 35, 47-58 (2001).

9. J. V. Carstensen, and T. Magnin, **Int. J. Fatigue**, 23, 195-200 (2001).

10. Z. Shi, Crack Analysis in Structural Concrete: Theory and Applications, p 172-78, Butterworth-Heinemann, Burlington, MA (2009).

11. Z. Shi, Y. Nakamura, and M. Nakano, **Int. J. Fatigue**, 33, 1140-50 (2011).

12. Z. Shi, M. Ohtsu, M. Suzuki, and Y. Hibino, **J. Structural Engineering**, 127, 1085-91 (2001).

13. Z. Shi, **J. Structural Engineering**, 130, 1738-47 (2004).

14. Z. Shi, CAIC-M1.FOR: <http://elsevierdirect.com/companion.jsp?ISBN=9780750684460>, the companion website of Crack Analysis in Structural Concrete (Butterworth-Heinemann, 2009).

15. A. A. Komarovsky, Physics of Strength and Fracture Control: Adaptation of Engineering Materials and Structures, p 467-90, CRC Press, (2003).

16. T. Ekobori, Physics and Mechanics of Fracture and Strength of Solids, **Metallurgiya** (1971; in Russian).

Fatigue of Materials III: Advances and Emergences in Understanding
Edited by: T.S. Srivatsan, M. Ashraf Imam, and R. Srinivasan
TMS (The Minerals, Metals & Materials Society), 2014

STRAIN ENERGY DIAGRAM FOR CHARACTERISING FATIGUE BEHAVIOUR IN STRUCTURAL MEMBERS AND MACHINE PARTS SUBJECTED TO REPEATED LOADS

Zihai Shi[1], Yukari Nakamura[1], Masaaki Nakano[1]

[1]Research and Development Centre
Nippon Koei Co., Ltd.
2304 Inarihara, Tsukuba
Ibaraki 300-1259, **Japan**
Email: a4739@n-koei.co.jp (SHI)

Abstract

According to the multistage strength degradation theory which has recently emerged from studies on the material and structural behaviour of concrete, fatigue is caused by the sporadic sudden change of cracking behaviour in a system under cyclic loading, leading to intermittent strength reduction of the system and its eventual failure. Here we report a new finding on the waveform variation of the maximum strain energy along the axis of load cycles in a structural member or machine part under fatigue. A strain energy diagram is calculated based on the system's response to the maximum load at any given time. The diagram shows the accumulation and release of total strain energy of a system at its ultimate strength, synchronising with the cycle of strength degradation. Therefore, this diagram provides real-time information on how a system responds to cyclic loading, which is of great importance to both fundamental research and design practice.

1. Introduction

The multistage strength degradation theory, which has recently emerged from studies on the material and structural behaviour of concrete [1, 2], provides a clear description of the fatigue mechanism that has been the focus of extensive research since the early nineteenth century [3-9]. According to this theory, fatigue is caused by the sporadic sudden change of cracking behaviour in a system under cyclic loading, leading to intermittent strength reduction of the system and its eventual failure. Here we report a new finding on the waveform variation of a strain energy diagram in association with the multistage strength degradation process. This diagram has novel features which are of considerable engineering interest.

Figure 1 shows schematic illustrations of the multistage strength degradation process of a system under fatigue and the corresponding strain energy diagram. The wavy curves of the diagram immediately reminded us of an electrocardiogram for diagnosing heart conditions, suggesting that the diagram could be useful for studying fatigue. As seen from **Figure 1(b)**, a strain energy diagram (SED) shows the accumulation and release of the total strain energy of the system at its ultimate strength, synchronising with the cycle of strength degradation. Therefore, this diagram provides real-time information on how a structural member or a machine part responds to cyclic loading, which is of great importance to both fundamental research and design practice. The basic terms that characterise the cyclic variation of an SED can be found in **Figure 1(b)**: the initial energy E_{ini} and the maximum energy E_{max}; the energy increase ΔE_i and the energy decrease ΔE_d; the energy increase rate r_i (= $\Delta E_i/E_{ini}$) and the energy decrease rate r_d (= $\Delta E_d/E_{max}$); and the region of energy accumulation R_{EA} and the region of energy release R_{ER}.

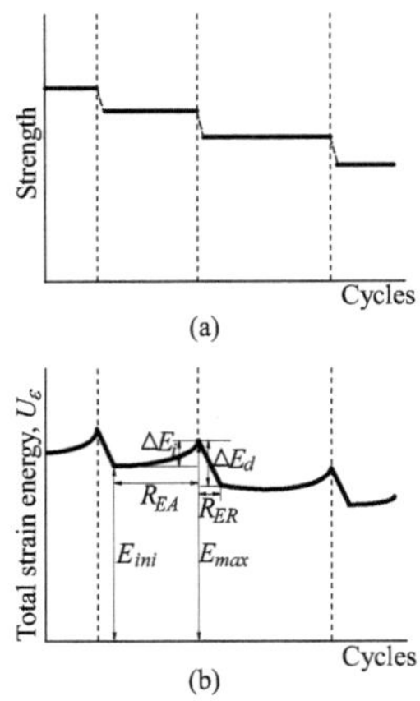

Figure 1. **Conceptual illustrations of multistage strength degradation and strain energy diagram: (a) a multistage strength degradation process under cyclic loading; (b) a strain energy diagram in association with the process of multistage strength degradation.**

2. The Energy Principles

To understand the theoretical basis of the SED, the energy transaction in the fracture process needs to be explained using Griffith's fracture theory [7, 10], whose innovative concept of introducing fracture energy into the study of cracked materials laid a solid foundation for the later development of fracture mechanics. First, the related theories are outlined below [11, 12].

2.1 Griffith's Fracture Theory

Griffith derived his fracture criterion by considering the problem shown in **Figure 2**, where an infinite plate of unit thickness that contains a crack of length $2a$ is subjected to uniform tensile stress σ. He first obtained the decrease of strain energy in the plate due to the introduction of the crack based on the stress analysis of Inglis [13], as

$$U_a = -\frac{\pi a^2 \sigma^2}{E} \text{ (plane stress)} \tag{1a}$$

$$U_a = -\frac{1-\nu^2}{E}\pi a^2 \sigma^2 \text{ (plane strain)} \tag{1b}$$

where E is the modulus of elasticity, and ν is Poisson's ratio. Griffith then introduced "the fundamental conception of the new theory", i.e., the surface energy, which is the fracture energy required to form the open crack, as

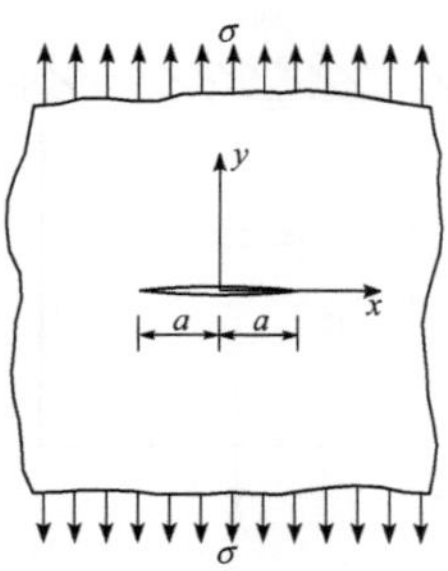

Figure 2. An infinite plate with a central crack subjected to tension.

$$U_s = 4a\gamma_s \tag{2}$$

where γ_s is the surface energy per unit area. Note that in **Eq. 2** the crack surface is $4a$ because a crack has two surfaces. According to the Griffith energy criterion, crack growth is a natural course of energy transfer between the strain energy of an elastic body and the fracture energy required for creating a new crack surface. The transitional equilibrium condition for stable crack growth requires the following condition to be satisfied:

$$\frac{d}{dA}(U_a + U_s) = \frac{1}{2}\frac{d}{da}(U_a + U_s) = 0 \tag{3}$$

Therefore, the critical load for crack propagation is obtained as

$$\sigma_c = \sqrt{\frac{2E\gamma_s}{\pi a}} \quad \text{(plane stress)} \tag{4a}$$

$$\sigma_c = \sqrt{\frac{2E\gamma_s}{(1-\nu^2)\pi a}} \quad \text{(plane strain)} \tag{4b}$$

Equation 4 is referred to as the Griffith fracture criterion, which is valid for ideally brittle materials.

2.2 The Energy Release Rate G

The Griffith energy concept is of a global character: for a crack to propagate, the energy stored in the system must be sufficient to overcome the fracture energy of the material. Based on Griffith's fracture theory, Irwin generalised the concept by defining an energy release rate G, which is a measure of the energy available for a unit extension of the crack and represents "the force tending to cause crack extension" [8]. **Figure 3** shows a cracked plate under a given load P, which causes the crack to extend by an amount dA. Since the load is fixed at P, the system is under load control. Obviously, the fracture energy dissipated in propagating the crack originates from the work done by the external force $Pd\Delta$ and the released elastic strain energy $-dU$ in the system, i.e.,

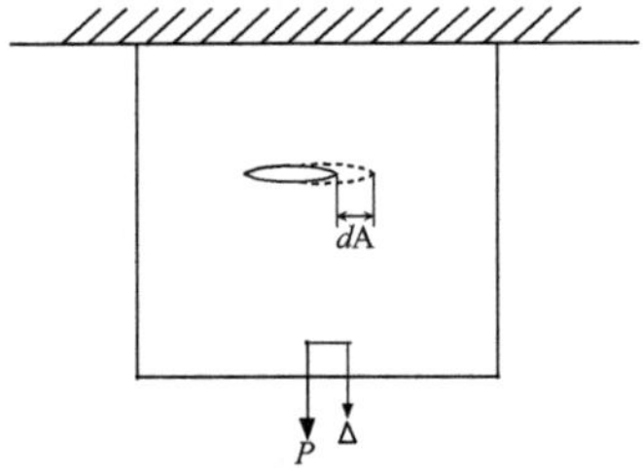

Figure 3. A cracked plate under a load *P*.

$$G = P\frac{d\Delta}{dA} - \frac{dU}{dA} \tag{5}$$

The strain energy of the elastic system is obtained as

$$U = \int_0^{\Delta} P d\Delta = \frac{P\Delta}{2} = \frac{CP^2}{2} \tag{6}$$

where the relation $\Delta = CP$ is employed; C is the compliance of the system, which is the inverse of the plate stiffness. Note that

$$\frac{d}{dA} = \frac{\partial}{\partial A} + \frac{dP}{dA}\frac{\partial}{\partial P} \tag{7}$$

After substituting **Eq. 6** and $\Delta = CP$ into **Eq. 5**, the above derivative operations are carried out and the energy release rate is derived as

$$G = \frac{P^2}{2}\frac{\partial C}{\partial A} \tag{8}$$

It is known that **Eq. 8** can also be obtained under the condition of displacement control. Theoretically, if the derivative of the compliance of a test specimen with respect to the crack length can be derived experimentally or numerically by an FE solution, then the energy release rate can be calculated from **Eq. 8**.

The energy release rate G can also be obtained from the stress intensity factor K. For mode I fracture the simple relationship between the two is given by

$$G = \frac{K_I^2}{E} \quad \text{(plane stress)} \tag{9a}$$

$$G = (1-\nu^2)\frac{K_I^2}{E} \quad \text{(plane strain)} \tag{9b}$$

where K_I is the mode I stress intensity factor.

According to Griffith's fracture theory, the decrease of the total strain energy ΔE_d in **Figure 1** can be obtained by integrating the energy release rate G over the crack increment Δa, i.e.,

$$\Delta E_d = \int_0^{\Delta a} G da \tag{10}$$

From **Eq. 9** it is clear that, just like the stress intensity factor K, the energy release rate G is also a function of the external load, the crack size and the geometry of the problem.

3. Multiple Cracking Modes in Fatigue Crack Growth

3.1 Crack Growth under Cyclic Loading

In **Figure 3**, after a small extension of the crack which releases both the critical stress from the crack tip and the excess strain energy from the system, the crack stops growing under the constant load P. In order to extend the crack further, the strain energy stored in the plate has to be increased by enlarging the load first.

To understand the mechanism of fatigue, one must remember that all materials contain a population of initial flaws and defects or fine cracks, which is one of the essential elements of Griffith's fracture theory. In a multi-crack situation, every crack possesses three possibilities of motion including crack extension, crack arrest and crack closure, and combinations of these modes among all the cracks involved will lead to a great number of possibilities for cracks to propagate. Consider a two-crack problem as an example. **Figure 4** shows an infinite plate of unit thickness with two arbitrarily-positioned interacting cracks of lengths A1 and A2, respectively. The plate is stressed to allow crack propagation. **Figure 4** illustrates all the five potential crack patterns or cracking modes, such as the single-crack-propagation modes in (a), the simultaneous-crack-propagation mode in (b), and the crack-propagation-with-crack-closure modes in (c). Since material damage due to fracture is irreversible, the closure of a crack here simply means the closing of the crack surfaces under local compressive stresses, without reversing the damaged material properties.

Under cyclic loading, continuous crack growths involving multiple cracking modes of similar fracture properties provide a reasonable explanation for the gradually-occurring fracture process of a material under fatigue, which is fundamentally different from the material fracture caused by a monotonically increasing load. Replacing the single crack in **Figure 3** by two arbitrary cracks and the constant load by a cyclic load of the same amplitude *P*, such as shown in **Figure 5**, continuous crack growths under cyclic loading then become possible. To facilitate the discussion, it is assumed that when the cyclic load reaches its maximum amplitude in the first load cycle, the Griffith fracture condition is met and the first crack in **Figure 5(a)** extends slightly. Obviously, the releasing of the critical stress from the tip of this crack makes continuous growth of the crack in the same cracking mode in the next immediate load cycle unlikely. However, following stress redistribution the likelihood of growth by the second crack may greatly increase, as the maximum strain energy in the system under the cyclic loading basically remains the same as when the first crack propagates. Hence, it is logical that the subsequent crack growth will take place in a cracking mode involving the second crack. The potential cracking modes for the present case are shown in **Figure 5(b)**, which include the single-crack-propagation mode of the second crack and the other interactive growth modes of the two cracks. Note that the real cracking mode should represent the minimum energy release rate *G* among all the possible modes. Clearly, following the second round of crack growth, the third round will take

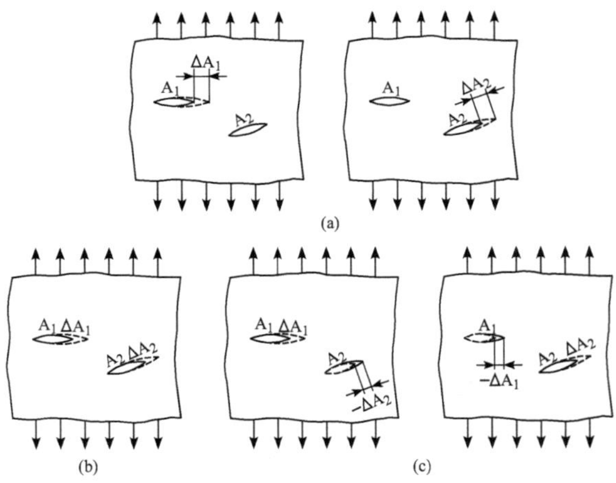

Figure 4. **A cracked plate with two arbitrary cracks subjected to tension and five potential cracking modes: (a) single-crack-propagation modes; (b) a simultaneous-crack-propagation mode; (c) crack-propagation-with-crack-closure modes.**

place in a similar fashion and further rounds thereafter, leading to a slow but continuous fracture process under cyclic loading.

3.2 Numerical Formulations of Multi-Crack Problems in Concrete

In the FE formulation of a multi-crack problem, as the next-step cracking behaviour cannot be uniquely determined, multiple cracking modes have to be taken into account. The following formulations are based on the single-crack-propagation modes. It will be shown later that more complex cracking modes can be obtained by correcting a presumed cracking mode when it leads to invalid solutions.

Figure 6 illustrates two cracks of the mode-I type, crack A and crack B, where crack propagation is set in the direction normal to the tensile force at the tip of each fictitious crack. In formulating crack equations, subscripts a and b represent, respectively, crack A and crack B, and l stands for the limit value of a nodal force (the tensile strength of concrete times the surface area apportioned to a nodal point). Superscripts i, j and k denote the corresponding nodes at designated cracks. For clarity, the cohesive forces F and the crack-opening displacements (CODs) W of the inactive crack are marked by asterisks. Initially, crack A is assumed to be the sole propagating crack. Hence, the tensile force at its tip must reach the nodal force limit Q_{la}, given by

$$Q_{la} = CR_a \cdot P_a + \sum_{i=1}^{N} CI_{aa}^{i} F_a^{i} + \sum_{j=1}^{M} CI_{ab}^{j} F_b^{*j} \tag{11}$$

where N and M are the number of nodes inside each fictitious crack, respectively. Note that the tensile forces at the tip of crack A, CR_a, CI_{aa}^{i} and CI_{ab}^{j}, are due to a unit external load, a pair of unit cohesive

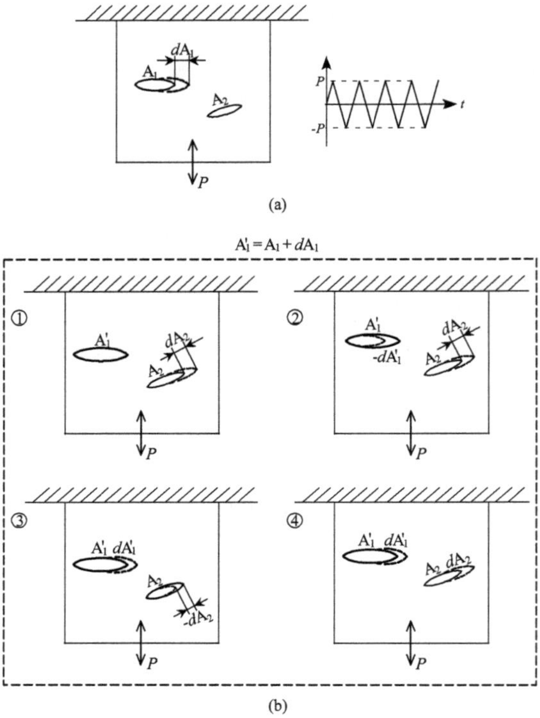

Figure 5. **Two interacting cracks under cyclic loading and potential cracking modes: (a) single crack propagation by the first crack; (b) potential cracking modes involving the second crack.**

forces at the i-th node of crack A, and a pair of unit cohesive forces at the j-th node of crack B, respectively. The external load P_a is the required load for propagating crack A, while crack B remains inactive. It should be noted that the tip force components due to the cohesive forces of crack B in **Eq. 11** represent crack interaction.

The CODs along the two fictitious cracks are obtained as

$$W_a^i = BK_a^i \cdot P_a + \sum_{k=1}^{N} AK_{aa}^{ik} F_a^k + \sum_{j=1}^{M} AK_{ab}^{ij} F_b^{*j} \tag{12}$$

$$W_b^{*j} = BK_b^j \cdot P_a + \sum_{i=1}^{N} AK_{ba}^{ji} F_a^i + \sum_{k=1}^{M} AK_{bb}^{jk} F_b^{*k} \tag{13}$$

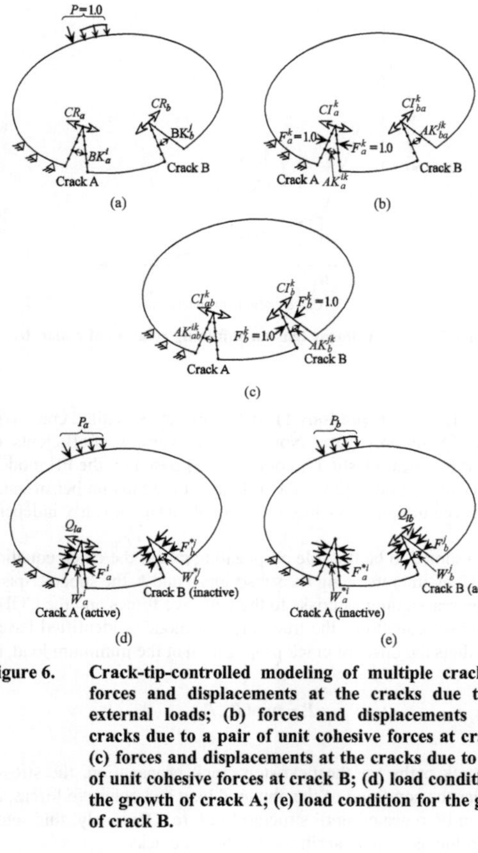

Figure 6. Crack-tip-controlled modeling of multiple cracks: (a) forces and displacements at the cracks due to unit external loads; (b) forces and displacements at the cracks due to a pair of unit cohesive forces at crack A; (c) forces and displacements at the cracks due to a pair of unit cohesive forces at crack B; (d) load condition for the growth of crack A; (e) load condition for the growth of crack B.

where $i = 1,\ldots, N$ and $j = 1,\ldots, M$. Here, the compliances $BK_a^{\,i}$ at crack A and $BK_b^{\,j}$ at crack B are due to the external load. The influence coefficients $AK_{aa}^{\,ik}$ and $AK_{ab}^{\,ij}$ are the CODs at the i-th node of crack A due to a pair of unit cohesive forces at the k-th node of crack A, and a pair of unit cohesive forces at the j-th node of crack B, respectively. Similarly, the influence coefficients $AK_{ba}^{\,ji}$ and $AK_{bb}^{\,jk}$ represent the CODs at the j-th node of crack B due to a pair of unit cohesive forces at the i-th node of crack A, and a pair of unit cohesive forces at the k-th node of crack B, respectively. According to the reciprocity theorem, $AK_{aa}^{\,ik} = AK_{aa}^{\,ki}$, $AK_{bb}^{\,jk} = AK_{bb}^{\,kj}$, and $AK_{ab}^{\,ij} = AK_{ba}^{\,ji}$.

Finally, imposing the tension-softening law of concrete (**Figure 7**) along each fictitious crack leads to

$$F_a^i = f(W_a^i) \tag{14}$$

$$F_b^{*j} = f(W_b^{*j}) \tag{15}$$

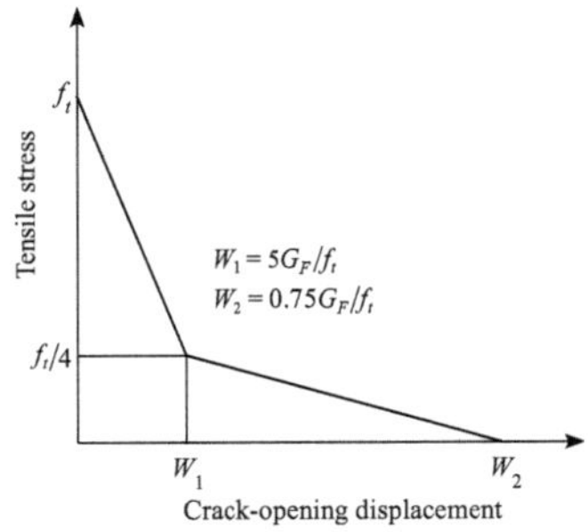

Figure 7. **Bilinear tension-softening relation of concrete.**

where $i = 1,\ldots, N$ and $j = 1,\ldots, M$. **Equations 11** to **15** form the so-called crack equations, stipulating the conditions for crack A to propagate. Note that the various coefficients employed in these equations can be obtained by linear elastic FE computations based on the FE models shown in **Figure 6(a)-(c).** With the number of equations ($2N + 2M + 1$) matching the number of unknowns ($2N + 2M + 1$), the problem can be solved uniquely because these equations are linearly independent.

When crack B is assumed to be the sole propagating crack, the crack equations can be obtained readily from **Eqs. 11** to **15** by interchanging the subscripts a and b, the superscripts i and j, the number of nodes N and M, and by reassigning asterisks to the cohesive forces and the CODs of crack A. Upon solving the two sets of crack equations, the true cracking mode is identified based on the minimum load criterion, which predicts the onset of crack propagation at the minimum load, that is,

$$P = \min(P_a, P_b) \tag{16}$$

After setting the true crack paths for the next-step crack propagation, the stress and displacement fields are calculated under the condition of the obtained load and cohesive forces, as shown in **Figure 6(d)-(e)**. This process can be repeated until structural failure. Obviously, this solution procedure can be readily extended to problems with an arbitrary number of cracks.

It should be noted that the numerical results thus obtained need to be checked carefully to eliminate invalid solutions that will be encountered when an assumed cracking mode is irrelevant to the problem. In a situation like this, the crack tip is readjusted by releasing or closing the tip nodes and the problem is recalculated. By relocating the crack-tip position of an inactive crack, other cracking modes with geometrically admissible strain fields may emerge, which include simultaneous propagations of several cracks, and crack growth accompanied by crack closure. For further details of the solution procedures and how to extend them to solve a mixed-mode fracture problem, refer to the previous publications [14, 15].

4. Three Cases for Numerical Studies on the SED

Figures 8 to **11** show three experimental and numerical cases for the present study. These cases were selected from previous studies on notched concrete beams to obtain evidence for multistage strength degradation in structural members under fatigue [1, 2]. We begin with our pioneering experimental study shown in **Figures 8** and **9**, which involved a series of four-point bending tests on plain concrete beams with five notches of various sizes cut into the specimens to simulate multiple

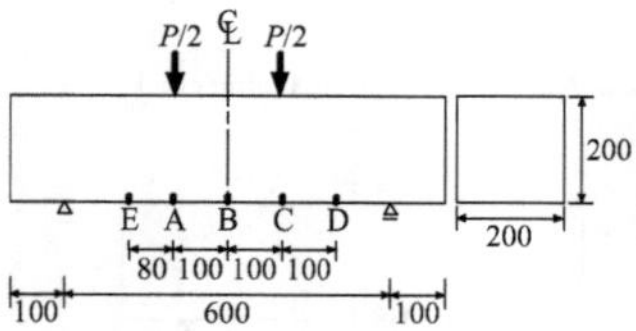

Notch size arrangements in tests

E	A	B	C	D
10	10	10	10	30,50,60
40,60	10	10	10	60
60	10	20,40	10	60

Unit: mm

Figure 8. Fracture tests on the load-carrying capacity of notched concrete beams under four-point bending with various arrangements of notch sizes.

Figure 9. Photo of a test specimen fractured at notch E.

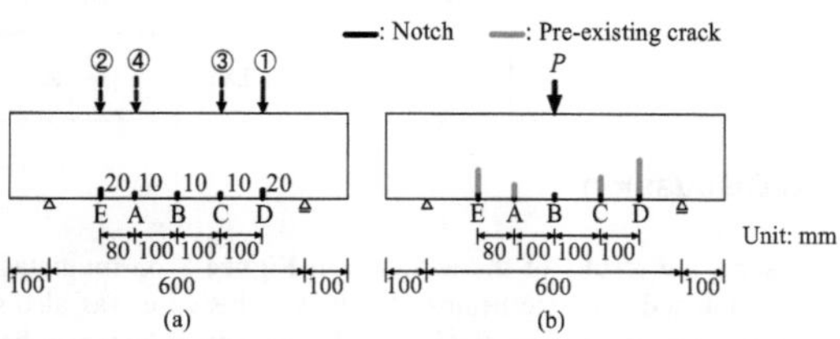

Figure 10. Numerical studies on the load-carrying capacity of notched concrete beams with cracks propagated by sequential loads (Type 1): (a) crack growths under sequential loading; (b) determining the maximum load under three-point bending with pre-existing cracks.

cracks and their growths under repeated loading. The material properties of the test specimens are shown in **Table I**: elastic modulus E, Poisson's ratio ν, compressive strength f_c, tensile strength f_t, and fracture energy G_F. We obtained a multistage strength degradation relation by increasing the notch sizes and carrying out fracture tests to investigate how the maximum loads decreased. To verify this relation based on realistic cracking behaviours, **Figures 10** and **11** present two case studies in which crack propagation was simulated numerically. In these studies, cracks were first propagated under sequential loads, and then at each step of crack growth the maximum load under three-point bending was calculated, assuming the previously-obtained configuration of crack propagations at the given step as the initial boundary condition of the beam. Using these coordinated computational procedures, we were able to obtain two multistage strength degradation relations, each indicating a typical type of fatigue failure. The material properties used for numerical studies are shown in **Table II**. In the following, the SED of each case is presented and discussed, and its possible engineering applications are briefly considered.

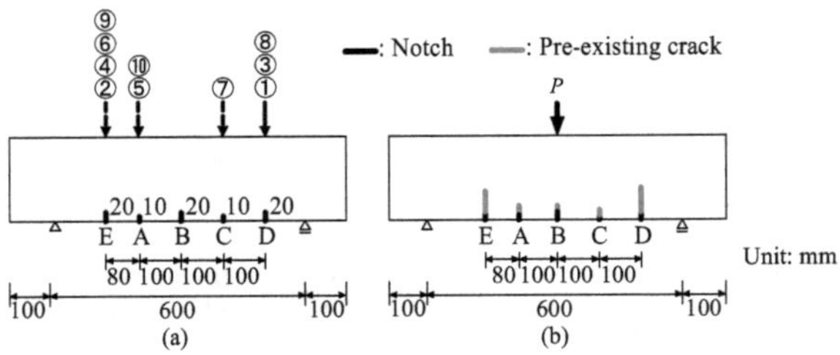

Figure 11. **Numerical studies on the load-carrying capacity of notched concrete beams with cracks propagated by sequential loads (Type 2): (a) crack growths under sequential loading; (b) determining the maximum load under three-point bending with pre-existing cracks.**

Table I. Material properties of concrete used in fracture tests.

E (kN/mm^2)	ν	f_c (kN/mm^2)	f_t (kN/mm^2)	G_F (N/mm)
29.2	0.18	34.3	3.17	0.142

Table II. Material properties of concrete used in numerical studies.

E (kN/mm^2)	ν	f_c (kN/mm^2)	f_t (kN/mm^2)	G_F (N/mm)
26.6	0.2	37.5	2.86	0.1

5. Results and Discussion

Figure 12(a) presents the results of the test case in **Figure 8** on multistage reductions of the load-carrying capacity in notched concrete beams. As shown, this case was also studied numerically using the crack analysis theories on multiple cracks and the agreement between the two results is good. Since detailed analyses of **Figure 12(a)** are presented elsewhere [1, 2], it is sufficient to point out here that the sudden transition of fracture behaviour which causes strength degradation occurs at notch D of 60 mm, notch E of 60 mm, and notch B of 30 mm, along the horizontal axis of notch size. The total strain energy of the beam at the maximum load is calculated from the work performed by the external loads, based on the load-displacement relations at the two load points shown in **Figure 13**. Based on **Figure 14**, the total strain energy, U_ε, is obtained as

$$U_\varepsilon = \int_0^{\Delta_m} P(\Delta) d\Delta \tag{17}$$

The following analysis is based on the numerical results of the SED shown in **Figure 12(b)**; the computational points in the diagram are numbered for convenience.

As shown in **Figure 12(b)**, there are three cycles of energy accumulation and release, i.e., the first cycle from Points 1 to 7 (E_{ini} = 1.61 N·m; E_{max} = 1.97 N·m; r_i = 22.4%; r_d = 20.8%), the second from 7 to 11 (E_{ini} = 1.56 N·m; E_{max} = 2.21 N·m; r_i = 41.7%; r_d = 37.6%), and the third from 11 to 14 (E_{ini} = 1.38 N·m; E_{max} = 1.47 N·m; r_i = 6.5%; r_d = 23.1%). In the regions of energy accumulation

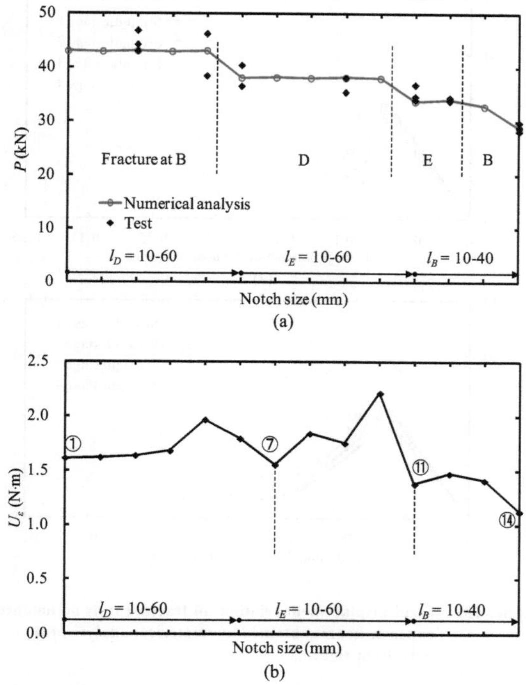

Figure 12. Results of experimental and numerical studies of fracture tests on notched concrete beams: (a) multistage strength degradation; (b) strain energy diagram based on numerical analysis.

(from Points 1 to 5, 7 to 10 and 11 to 12) the load-carrying capacity of the beam remains constant, as shown in **Figure 12(a)**. Bearing the Griffith energy-balance theory in mind, the rapid energy increments from Point 4 and Point 9 are remarkable, as if the beam is preparing itself with extra strain energy for the upcoming major transition in fracture behaviour. In fact, this is the result of the intense crack interaction that often occurs prior to the emergence of new cracking behaviours, hence increasing the system's nonlinearity and its strain energy as well. The energy release from Points 5 to 7, 10 to 11, and 12 to 14 indicates the onset of strength degradation and this coordination between the two events reveals their inherent relationship. Note that the temporary energy decrease at Point 9 in a region of energy accumulation is the result of the delicate crack interactions, which reduce the beam deformation near notch D.

Figures 15 to **18** present numerical results of the two case studies of **Figures 10** and **11**. Note that while in **Figure 15** strength degradation occurs when a propagating crack reaches a critical length, in **Figure 17** it is triggered by simultaneous propagations of crack B amid growth of the surrounding cracks. **Figures 16** and **18** present the load-displacement relations of each case, respectively. Here we focus on the unique features of the two SEDs; for a detailed discussion of the strength diagrams, refer to the previous study [2]. As shown in **Figure 15(b)**, three cycles of energy transactions are identified: from points 1 to 7 (E_{ini} = 1.34 N·m; E_{max} = 1.64 N·m; r_i = 22.4%; r_d = 3.0%), 7 to 13 (E_{ini} = 1.59 N·m; E_{max} = 1.83 N·m; r_i = 15.1%; r_d = 5.5%), and 13 to 16 (E_{ini} = 1.73 N·m; E_{max} = 1.88 N·m; r_i = 8.7%; r_d

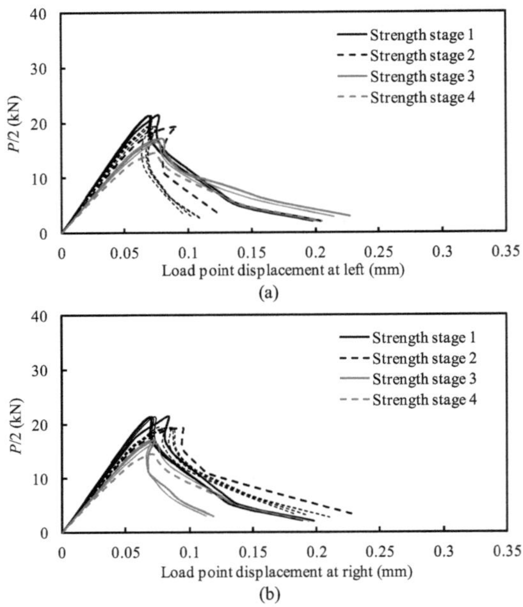

Figure 13. **Load-displacement relations of fracture tests on notched concrete beams obtained by numerical analysis: (a) at left; (b) at right.**

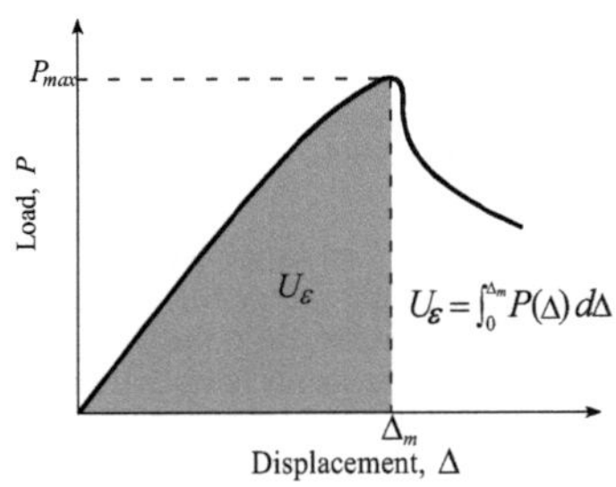

Figure 14. **Integration of the load-displacement relation to obtain the total strain energy at the maximum load.**

= 10.6%). The fourth is incomplete, as its last point is obviously still in the stage of energy accumulation. The temporary change of cracking behaviour at point 14 in the third cycle, as shown in **Figure 15(a),** is worth commenting on. The fact that point 14 is in the region of energy accumulation explains why the ultimate strength of the beam remains constant at this point.

Now turning to the SED shown in **Figure 17(b)**, the following data are sufficient for characterising the main features of its four cycles: points 1 to 4 (E_{ini} = 1.12 N·m; E_{max} = 1.14 N·m; r_i = 1.8%; r_d = 11.4%); 4 to 10 (E_{ini} = 1.01 N·m; E_{max} = 1.12 N·m; r_i = 10.9%; r_d = 22.3%); 10 to 12 (E_{ini} = 0.87 N·m; E_{max} = 0.9 N·m; r_i = 3.4%; r_d = 11.1%); and 12 to 16 (E_{ini} = 0.8 N·m; E_{max} = 0.86 N·m; r_i =

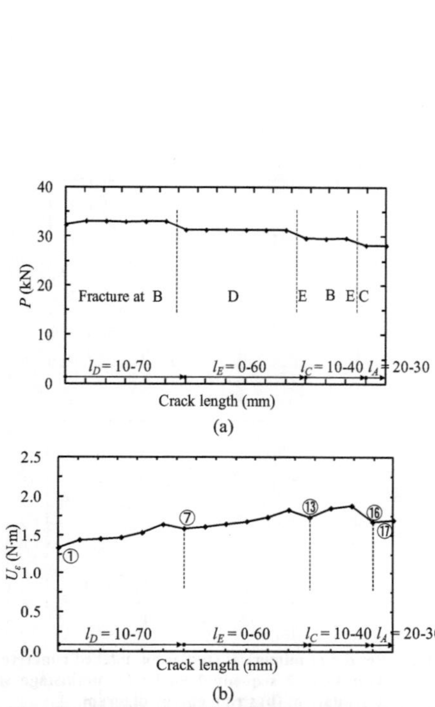

Figure 15. **Results of numerical studies on notched concrete beams with Type 1 sequential loads: (a) multistage strength degradation; (b) strain energy diagram.**

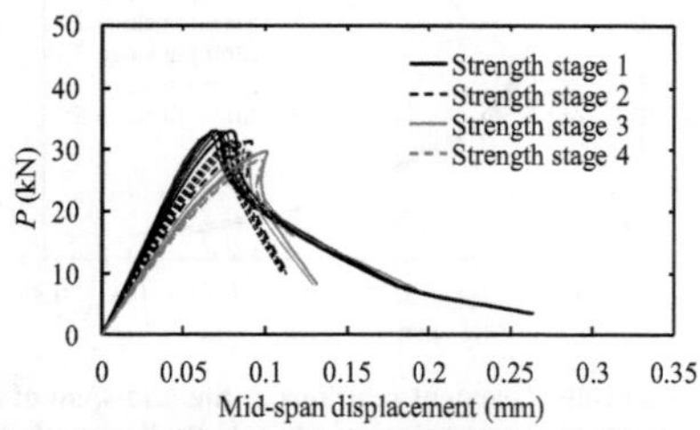

Figure 16. **Load-displacement relations at the mid-span of notched concrete beams in numerical studies with Type 1 sequential loads.**

7.5%; r_d = 22.1%). Note that, among the three cases discussed so far, this is the only case in which its strength degradation does not involve a change of failure mode.

Although the horizontal axes in **Figures 12, 15** and **17** are shown in either notch size or crack length, for a structural member or a machine part under cyclic loading these strength and energy relations should be calculated based on the number of load cycles experienced by the system. Therefore, in practice the horizontal axis of an SED represents the number of load cycles. One possible application of the diagram is for evaluating the safety of an engineering system that has sustained crack-related damage under repeated loading. Since in theory the SED for that system can be established by numerical analysis and tests, the diagram provides a tool for studying the general structural or mechanical behaviour of the system related to fatigue, based on the calculation and study of its basic terms, i.e., E_{ini}, E_{max}, r_i and r_d. The diagram can also be used for assessing the present state of the system under fatigue, based on the number of load cycles experienced and the crack damage sustained. The obtained energy level represents the maximum strain energy the system can sustain at

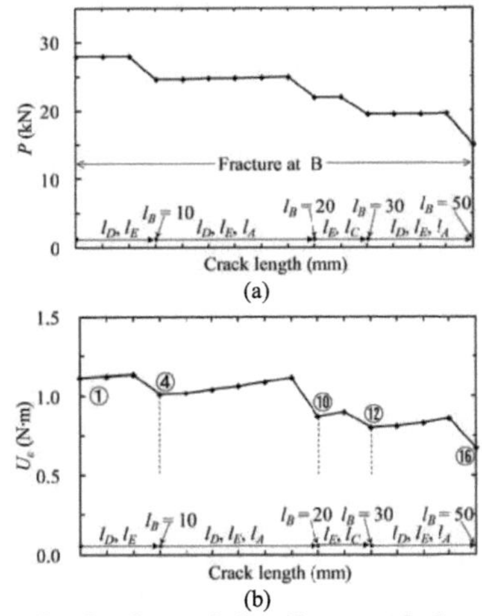

Figure 17. Results of numerical studies on notched concrete beams with Type 2 sequential loads: (a) multistage strength degradation; (b) strain energy diagram.

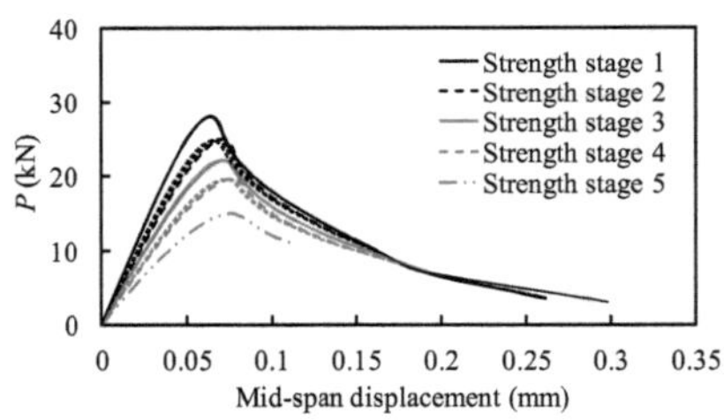

Figure 18. Load-displacement relations at the mid-span of notched concrete beams in numerical studies with Type 2 sequential loads.

the given time, and the ratio of the actual strain energy under cyclic loading to this maximum value may provide a general representation of the state of fatigue damage sustained by the system: the larger the ratio, the greater the damage. Finally, for fatigue design this diagram may provide a new theory for evaluating the system's performance under cyclic loading.

6. Summary

Failure due to fatigue, i.e., repeated loading, is multidisciplinary and is the most common cause of engineering failure. Despite the enormous progress made in developing fatigue behaviour models and design principles over more than 160 years, the mechanism of fatigue is still not fully understood. The recently-developed multistage strength degradation theory provides a clear description, stating that fatigue is caused by the sporadic sudden change of cracking behaviour in a material, which can lead to intermittent strength reduction of a system under cyclic loading. This paper presents a new finding on the waveform variation of a strain energy diagram in association with the multistage strength degradation process. A strain energy diagram is calculated based on the response of a structural component or a machine part with a prior loading history to the maximum external load. The diagram shows the accumulation and release of the total strain energy of a system at its ultimate strength, synchronising with the cycle of strength degradation along the axis of load cycles. Therefore, this diagram provides real-time information on how a system responds to cyclic loading, which is of great importance to both fundamental research and design practice.

References

1. Z. Shi, Crack Analysis in Structural Concrete: Theory and Applications, Butterworth-Heinemann, Burlington, MA, 172-78 (2009).

2. Z. Shi, Y. Nakamura, and M. Nakano, **Int. J. Fatigue** 33, 1140-50 (2011).

3. W. A. J. Albert, Archive für Mineralogie, Geognosie, Bergbau und Hüttenkunde 10, 215-34 (1838).

4. A. Wöhler, Zeitschrift für Bauwesen 10 (1860); English summary: Engineering 4, 160-61 (1867).

5. W. Fairbairn, **Phil. Trans. Roy. Soc.**, 154, 311 (1864).

6. J. A. Ewing, and J. C. Humfrey, **Phil. Trans. Roy. Soc.,** A200, 241-50 (1903).

7. A. A. Griffith, **Phil. Trans. Roy. Soc.**, A221, 163-97 (1921).

8. G. R. Irwin**, Journal of Applied Mechanics**, 24, 361-64 (1957).

9. P. C. Paris, M. P. Gomez, and W. P. Anderson, **The Trend in Engineering**, 13, 9-14 (1961).

10. A. A. Griffith, Proceeding of First International Congress for Applied Mechanics, C. B. Biezeno and J. M. Burgers, Eds., 55-63 (1925).

11. T. L. Anderson, Fracture Mechanics: Fundamentals and Applications, CRC Press, Boca Raton, FL, p 29-38 (2005).

12. R. J. Sanford, Principles of Fracture Mechanics, Prentice Hall, p 236-50 (2003).

13. C. E. Inglis, Trans. Inst. Naval Architects, 55, 219-30 (1913).

14. Z. Shi, M. Ohtsu, M. Suzuki, Y. Hibino, **J. Struct. Eng.**, 127, 1085-91 (2001).

15. Z. Shi**, J. Struct. Eng.**, 130, 1738-47 (2004).

FATIGUE of MATERIALS III
ADVANCES and EMERGENCES in UNDERSTANDING

Ferrous Materials

Fatigue of Materials III: Advances and Emergences in Understanding
Edited by: T.S. Srivatsan, M. Ashraf Imam, and R. Srinivasan
TMS (The Minerals, Metals & Materials Society), 2014

CYCLIC STRAIN RESISTANCE, DEFORMATION AND FRACTURE BEHAVIOR OF A NOVEL ALLOY STEEL

K. Manigandan and T. S. Srivatsan
Division of Materials Science and Engineering
Department of Mechanical Engineering
The University of Akron
Akron, Ohio 44325-3903
E-Mail: tsrivatsan@uakron.edu

V. K. Vasudevan, D. Tammana and B. Poorbangi
School of Dynamic Systems
Materials Science and Engineering Program
University of Cincinnati
Cincinnati, Ohio 45221-0072
E-Mail: vijay.vasudevan@uc.edu

Abstract

In this paper, the results of a study on microstructural influences on cyclic strain response, deformation and fracture behavior of an alloy steel is presented Cyclic strain resistance exhibited a linear trend for the variation of both elastic strain amplitude with reversals-to-failure, and plastic strain amplitude with reversals-to-failure. Fracture morphology was observed to be the same at the macroscopic level over the entire range of cyclic strain amplitudes examined. However, at the fine microscopic level this alloy steel revealed fracture to be mixed-mode with features reminiscent of "locally" ductile and brittle failure mechanisms. The mechanisms governing strain response at the fine microscopic level, resultant fatigue life, and final fracture behavior are presented and discussed in light of the mutually interactive influences of intrinsic microstructural effects, deformation characteristics of the microstructural constituents during fully-reversed strain cycling, magnitude of cyclic strain amplitude, and resultant fatigue life.

Key Words: alloy steel, microstructure, tensile properties, fatigue

1. Introduction

The last four decades, i.e., since the early 1980's, manufacturers of all kinds of steels, spanning the domains of carbon steels, mild steel, stainless steels, alloy steels, tool steel and specialty steel have attempted to successfully engineer the use of these steels for a spectrum of applications spanning the domains of performance-critical and non-performance critical. This in turn has provided them with not only the required incentive but also the impetus to shift their attention and action towards the production, processing and use of new and improved high strength steels. The newer class of high strength steels is noted for their ability to offer an attractive combination of high tensile strength, acceptable fracture toughness, improved weldability coupled with acceptable performance in environments spanning a range of aggressiveness, to include both aqueous and gaseous, when compared one-on-one with the mild steel counterpart [1-10]. During the same time period production and processing of the family of new and improved high strength steels did culminate from progressive advancement of the processing and manufacturing techniques in the steel industry, particularly the domain of thermo-mechanical processing (TMP). In this connection, a control of heating and subsequent rolling process during production of the steel can result in the formation of a fine grain size end product.

The mechanical properties of the family of new and improved family of high strength steels are governed by the conjoint and mutually interactive influences of chemical composition, processing history, intrinsic microstructural features, temperature of operation, loading rate, and even potential constraints at the crack tip [11-20]. Conditions that are often favorable for minimizing crack tip constraints are also conducive for enhancing the ability of the candidate steel to be receptive to plastic deformation. This facilitates an overall improvement in fracture toughness coupled with an increased tolerance to fatigue damage at high temperatures. In more recent years, several studies have made an attempt to show the factors, ranging from: (i) an increase in part thickness, (ii) incorporation of engineering notches, (iii) increase in loading rate, and (iv) a lowering of the operating temperature, are often detrimental to both ductility and toughness of the steel [21-25].

In this paper, we present and discuss the results of a study aimed at understanding the influence of intrinsic microstructural effects on cyclic strain amplitude-controlled response; fatigue life and final fracture behavior of the high strength alloy steel Tenax^{TM} 310. The likely mechanisms governing cyclic strain resistance and resultant fatigue life, and final fracture behavior, are presented in light of the conjoint and mutually interactive influences of intrinsic microstructural effects, magnitude of cyclic strain amplitude, deformation characteristics of the microstructural constituents, and overall fatigue life.

2. Material

The high strength steel chosen for this study was Tenax^{TM} 310. This steel was provided by Carpenter Technology, Inc (based in Reading, PA, USA). The chemical composition (in weight percent) of the high strength alloy steel is given in **Table 1.** The presence of carbon in substantial amounts in this steel provides solid solution strengthening besides increasing hardenability through the formation, presence and distribution of alloy carbides. Presence of the elements chromium (Cr) and molybdenum (Mo) results in the formation and presence of the second-phase particles chromium carbide (Cr_2C_3) and molybdenum carbide (Mo_2C), which are well dispersed through the microstructure. The carbide particles distributed through the microstructure contribute to enhancing strength of the steel matrix [26, 27]. The presence and distribution of a substantial number of such carbide particles in the microstructure is detrimental to both ductility and fracture toughness. This can be attributed to a 'local' increase in the number of sites that are easily susceptible to the early initiation of fine microscopic cracks. Presence of the element nickel (Ni) in the alloy steel helps in lowering the transition temperature thereby making the steel appropriate for use at low service temperatures. This steel, i.e.,Tenax^{TM} 310, also contains an appropriate amount of molybdenum, whose purpose is to refine the grain size besides forming molybdenum carbide (Mo_2C) and its indirect influence in contributing to enhancing both strength and impact toughness. This alloy steel was produced using the method of Arc VAR [VAR: Vacuum Arc Remelting); with Arc in this case being an initial arc-melt operation].

Table 1: Nominal chemical composition of Tenax ^{TM}310 (in weight percent)

Material	C	Mn	Si	P	S	Cr	Ni	Mo	Cu
Tenax ^{TM}310	0.401	0.63	2.03	0.004	<0.0005	1.26	3.76	1.01	0.53

3. Experimental Procedures

3.1 Characterization of As-received Microstructure

A sample of the chosen high strength steel Tenax^{TM}310 was prepared very much in conformance with the standard procedures used for metallographic preparation of metal samples. The as-polished sample was then etched using nital reagent, i.e., a solution mixture of nitric acid in methanol. The polished and etched sample was subsequently examined in an optical microscope, at low magnifications, and photographed using standard bright field illumination technique.

Transmission electron microscopy (TEM) observations were used to investigate grain size, grain distribution, presence, size and distribution of the second-phase carbide particles and other fine intrinsic features in the microstructure of the as-provided material. Samples 0.5 mm to 0.6 mm in thickness were sliced from both the undeformed section [i.e. grip section] and deformed section (i.e. gage length) of the alloy steel test sample. From these slices were punched out thin discs having an effective diameter of 3-mm and a thickness of around 100 μm. The circular discs, referred to henceforth as samples, were dimpled using a dimpler machine [Model: FISCHIONE] and a diamond solution gradually down to 1 μm size. Using the technique of dimpling, the thickness of the sample was effectively reduced to around 20-30 μm. Subsequently, ion milling was used for the final stage of sample preparation for purpose of observation in a transmission electron microscope (TEM). Argon gas ion-milling was used with the gun angle set at 14 degrees and in a chamber that was cooled by liquid nitrogen. The resultant polished sample, referred to as a "thin-foil", was observed in a transmission electron microscope (Model: Phillips) operating at 200 kV.

3.2 Preparation of Test Specimen

Cylindrical test specimens, conforming to specifications outlined in ASTM E8 [28], were precision machined from the chosen steel. The threaded test specimens measured 59 mm in overall length and 6.35 mm in diameter at the thread section. The gage section of the machined test specimen measured 12.5 mm in length and 3.175 mm in diameter. The length-to-diameter ratio of the fatigue test specimen was chosen to ensure that it would not buckle under conditions of fully-reversed [R_ε = minimum strain / maximum strain ($\sigma_{minimum}$ / $\sigma_{maximum}$)] total strain amplitude-controlled cyclic deformation. To minimize the effects and/or contributions arising from both surface irregularities and surface finish, final preparation of the test specimen surface was achieved by mechanically polishing the gage section of all specimens using progressively finer grades of silicon carbide impregnated emery paper to remove any and all of the circumferential scratches and surface machining marks.

3.3 Mechanical Testing (Tension plus Fatigue)

Tensile tests were performed up until failure on a fully-automated, closed-loop servohydraulic mechanical test machine [INSTRON Model 8500 plus] equipped with a 10,000 kgf (98 KN) load cell. The test specimens were deformed at a constant strain rate of 0.0001/s. An axial 12.5 mm gage length extensometer was attached to the test specimen at the gage section, using rubber bands, to provide a precise measurement of strain during uniaxial loading and resultant stretching of the test specimen. All of the stress and strain measurements, parallel to the load line, were recorded on a PC-based data acquisition system [DAS].

The fatigue tests were also performed on a fully-automated closed-loop servohydraulic structural test machine. The tests were conducted in the axial total strain amplitude control mode under fully-reversed, push-pull, tension-compression loading. The test machine was aligned prior to the initiation of each fatigue test very much in conformance with the guidelines put forth by the manufacturer (INSTRON Corporation, Canton, MA, USA). For each individual test the test machine was programmed to maintain a constant nominal strain rate of 0.001/s. The test common signal (i.e., strain function) was a triangular waveform, and the mean strain was zero. All of the total strain amplitude-controlled ($\Delta\varepsilon_T/2$) cyclic fatigue tests were initiated in tension. A 12.7 mm clip-on extensometer was attached to the test specimen at the gage section, using rubber bands, for the purpose of monitoring total strain-amplitude ($\Delta\varepsilon_T/2$) during fully-reversed strain amplitude-controlled fatigue tests such that the magnitude of negative strain equals the magnitude of positive strain ($R_\varepsilon = \varepsilon_{min.}/\varepsilon_{max} = -1$). The controlled variable is total strain amplitude ($\Delta\varepsilon_T/2$) and the cyclic strain-controlled tests were performed over a range of total strain amplitudes ($\Delta\varepsilon_T/2$). The extensometer was calibrated prior to the initiation of each fatigue test. The tests were conducted at room temperature (27° C) and in laboratory air environment (relative humidity of 55 percent). The stress and strain data for each fatigue test was recorded on a PC-based data acquisition system. The number of cycles-to-failure, or separation, is taken as fatigue life (N_f).

3.4 Analysis of Damage and Failure

The fracture surfaces of the cyclically deformed and failed fatigue test specimens of this alloy steel were comprehensively examined in a scanning electron microscope (SEM) to determine the following:

(i) the macroscopic fracture mode, and

(ii) characterize the fine scale topography, the nature of crack initiation, the extent and depth of early crack propagation, the extent and depth of stable crack propagation, and other intrinsic features on the fracture surface.

This was essential for establishing the microscopic mechanisms contributing to failure by fracture.

The distinction between the macroscopic mode and microscopic fracture mechanisms is based entirely on the magnification level at which the observations were made. The macroscopic mode referring to the overall nature of failure while the fine microscopic mechanisms being related to failure processes occurring at the "local or microscopic" level. This includes the following: (i) microscopic void formation, (ii) microscopic void growth and their eventual coalescence by way of impingement, and (iii) Nature, intensity and severity of the fine microscopic and macroscopic cracks dispersed through the fracture surface. The samples for observation in the scanning electron microscope (SEM) were obtained from the cyclically deformed and failed fatigue specimens by sectioning parallel to the fracture surface.

4. Results and Discussion

4.1 Initial Microstructure.

The optical microstructure of the chosen high strength steel is shown in **Figure 1.** The observed microstructure is typical of high strength steels in that it reveals a combination of carbon-rich and carbon-depleted regions. A higher carbon and alloy content in this high strength alloy steel results in a greater volume fraction of martensite in the carbon-rich regions. The presence and overall morphology of the martensite was much finer and intermingled with pockets of ferrite-rich regions, i.e., the carbon depleted region, in this high strength steel that was produced by vacuum arc re-melting (VAR). The presence of the hard, brittle and essentially elastic deforming martensite micro-constituent becomes a preferred site for high 'local' stress concentration during cyclic deformation. In addition to the presence of these two phases, or micro-constituents, i.e., ferrite and martensite, also dispersed through the microstructure were the alloy carbides. These particles result as a consequence of the alloy content in this high strength steel. The morphology of the martensite micro-constituent in the carbon-rich region is governed by a synergism of both chemical composition and processing technique, to include both primary processing and secondary processing that was used to engineer the starting material (Tenax^{TM} 310). These two phases along with the carbide particles exert an appreciable influence on cyclic fatigue response, fatigue life, nature of deformation and final fracture behavior, at the fine microscopic level, when specimen of the alloy steel is subject to cyclic loading.

Transmission electron microscopy observations of the thin foils taken from the undeformed alloy steel when viewed at low magnifications revealed a combination of elongated grains intermingled with the smaller grains. The grains were overall of varying size (**Figure 2a**). At the higher allowable magnifications of the TEM the observable and/or resolvable grains (**Figure 2b**) were oriented in the direction of mechanical deformation (**Figure 2c**).

4.2 Tensile Properties

The ambient temperature tensile properties of the candidate steel (Tenax^{TM}310) is summarized in **Table 2.** Results reported are the mean values based on duplicate tests. The elastic modulus of this steel is 194 GPa. The yield strength of this high strength alloy steel (Tenax^{TM} 310) is 1018 MPa and the ultimate tensile strength is 1225 MPa. The tensile strength is only marginally higher than the yield strength indicating the occurrence of slow degree of train hardening beyond yield. The ductility quantified by elongation over half-inch gage length was 14 percent. The reduction in test specimen cross-sectional area, another quantifiable measure of ductility, was 31 percent.

Table 2. A compilation of the room temperature tensile properties of the chosen steel

Material	Elastic Modulus		Yield Strength		UTS		Elong.	Reduction in Area
	Ksi	GPa	Ksi	MPa	Ksi	MPa	(%)	(%)
Tenax^{TM}310	**29747.2**	**205**	**147.7**	**1018.5**	**177.7**	**1225.4**	**13.4**	**31**

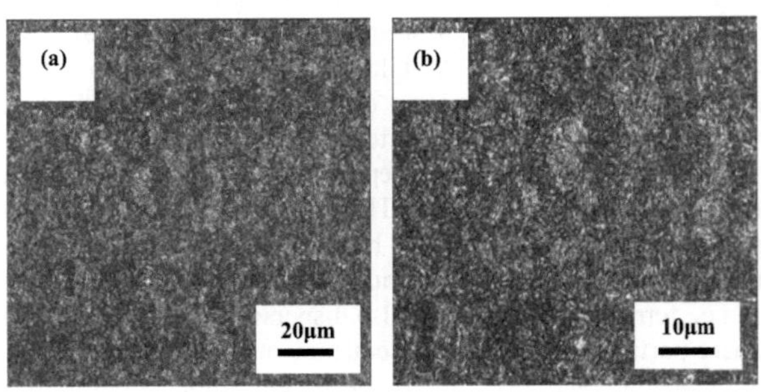

Figure 1. Optical micrographs showing the key micro-constituents present in alloy steel Tenax^{TM} 310 at two different magnifications.

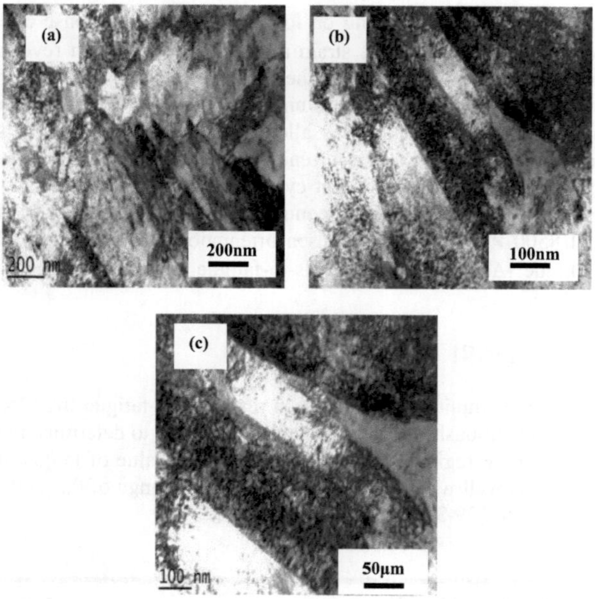

Figure 2. **Transmission electron micrographs showing the size and shape of grains and their orientation.**

4.3 Cyclic Strain-Controlled Fatigue Properties

A number of damaging processes are induced or caused by cyclic plastic strain. This often tends to affect not only the microstructure of this alloy steel, but also the cyclic strain resistance and resultant fatigue life [29-31]. The initiation and progressive growth of the damage through the microstructure of this alloy steel helps in relating the fatigue life (N_f) to cyclic strain amplitude [$\Delta\varepsilon$ /2] during fully-reversed [R_ε = -1] strain cycling. The relationship between plastic strain amplitude [$\Delta\varepsilon_P$ /2] and reversals-to- fatigue failure [$2N_f$], also known as the Coffin-Manson relationship, was determined for the LCF regime [$N_f < 10^4$ cycles] to be

$$[\Delta\varepsilon_P /2] = \varepsilon_f' [2N_f]^c \qquad [1]$$

The best straight line fit for the plastic strain amplitude [$\Delta\varepsilon_P$ /2] – reversals-to-fatigue life ($2N_f$) curve was obtained by linear regression analysis [29, 30]. An extrapolation of the best fit to the data points was made with the primary objective of determining the value of cyclic plastic strain amplitude ($\Delta\varepsilon_P$ /2) at 2 N_f = 1. This point in the plastic strain amplitude-fatigue life curve is referred to as the fatigue ductility coefficient [ε_f']. This coefficient provides a measure of cyclic ductility of the material. The slope of the plastic strain amplitude [$\Delta\varepsilon_P$ /2] -reversals-to-fatigue life ($2N_f$) curve gives the fatigue ductility exponent, **c**.

The effect of fully-reversed strain cycling on low cycle fatigue response of this alloy steel is shown in **Figure 3**. Variation of cyclic strain amplitude [$\Delta\varepsilon$ /2] with reversals-to-fatigue life [$2N_f$] curve can be viewed as an indication of the resistance offered by the microstructure of this alloy steel to both crack initiation and failure under conditions of fully-reversed strain cycling. Throughout strain-controlled testing of this alloy steel, over the range of strain amplitudes studied, there was no indication of the occurrence of specimen buckling. Based on width and height of the hysteresis loop for the different cycles, the plastic strain, elastic strain and stress amplitude values were recorded by the control unit of the servohydraulic test machine (INSTRON Model 8500 Plus) For the chosen orientation of this alloy steel the variation of elastic strain amplitude [$\Delta\varepsilon_e$ /2] with reversals-to-fatigue life [$2N_f$] is linear and the Basquin relationship is satisfied.

$$[\Delta\varepsilon_e/2] = \{\sigma_f' / E\} [2N_f]^b \qquad (2)$$

Variation of plastic strain amplitude [$\Delta\varepsilon_P$ /2] with reversals-to-fatigue life [$2N_f$] was also linear and the Coffin Manson relationship is satisfied and can be used to determine the fatigue behavior in the strain-controlled LCF regime [$N_f < 10^4$ cycles]. The value of fatigue ductility exponent [c] is 0.22. This value is well within the generally observed range of 0.2 to 0.7 for most metals and their alloy counterparts [29-31].

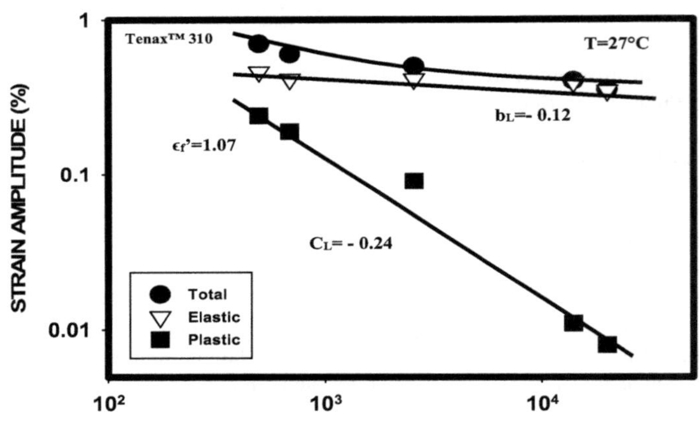

Figure 3. **Variation of cyclic strain amplitude ($\Delta\varepsilon/2$) with reversals- to-failure ($2N_f$) for alloy steel Tenax™310 at room temperature (27° C).**

4.4 Cyclic Stress Response

The cyclic stress response curves exemplify the variation of cyclic stress amplitude ($\Delta\sigma$ /2) with cycles (N), at a fixed strain amplitude [$\Delta\varepsilon_T$ /2]. These curves concurrently provide a lucid illustration of the path taken by this alloy steel to its final level of stress during fully-reversed total strain amplitude-controlled cyclic deformation. The curves also provide useful information pertaining to the overall cyclic stability of the chosen alloy steel. Stability of the intrinsic microstructural features, during fully-reversed strain cycling, in conjunction with an ability of the microstructure of this high alloy steel to distribute the plastic strain over the entire microstructural volume are factors that govern both the stress response and resultant fatigue life (N_f). Details of the nature of stress response, stress versus strain response and microscopic mechanisms governing the stress-strain response of this alloy steel can be found elsewhere [32].

4.5 Cyclic Fracture Behavior

At room temperature (27°C) fracture surfaces of the samples that were cyclically deformed at both high cyclic strain amplitude and resultant short fatigue life and low cyclic strain amplitude and concomitant long fatigue life reveal near similar topography at the macroscopic level. At the fine microscopic level the fracture surface features were found to vary with cyclic strain amplitude and resultant fatigue life. Representative fractographs of the fatigue fracture surface are shown in **Figures 4-7.**

4.5.1 High Cyclic Strain Amplitude (1.0 pct.) [Fatigue Life = 1,292 cycles]

At this value of total strain amplitude (1.00 pct.) and resultant fatigue life of 1,292 cycles macroscopic fracture of this alloy steel specimen was essentially normal to the far-field stress axis (**Figure 4a).** Macroscopic or low magnification, observation in the scanning electron microscope (SEM) revealed the region of crack initiation and early microscopic crack growth to be smooth and inlaid with an array of fine microscopic cracks intermingled with macroscopic cracks (**Figure 4 b**). The region of stable crack growth prior to overload fracture surface was microscopically rough (**Figure 4 c**). At higher allowable magnifications of the scanning electron microscope this region revealed a combination of microscopic cracks, voids and dimples. These fine microscopic features are reminiscent of both brittle and ductile failure mechanisms occurring at the fine microscopic level. The region of unstable crack growth prior to overload revealed a microscopically rough fracture surface with cracking along the grain boundaries and grain boundary triple junctions (**Figure 5a**). The region of overload revealed a sizeable population of dimples of varying size indicative of locally acting ductile failure mechanisms **(Figure 5b).**

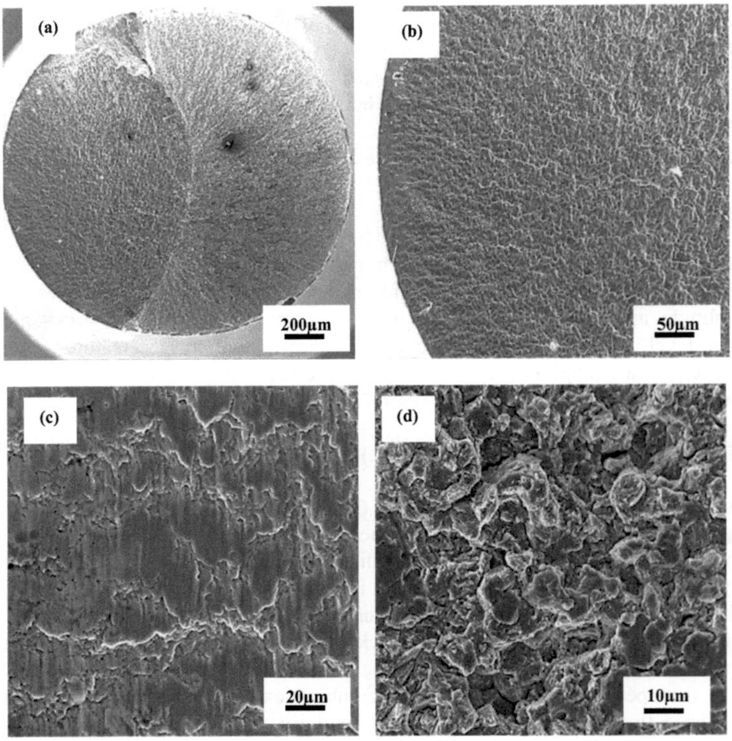

Figure 4. Scanning electron micrographs of the fatigue fracture surface of alloy steel Tenax™ 310 sample, cyclically deformed at a strain amplitude of 1.00%, with a fatigue life of 1,292 cycles, showing:
(a) Overall morphology of failure
(b) High magnification observation of (a), Microscopically rough
(c) Microscopic cracks in the region of crack initiation and early crack growth
(d) Region of stable crack growth: microscopically rough

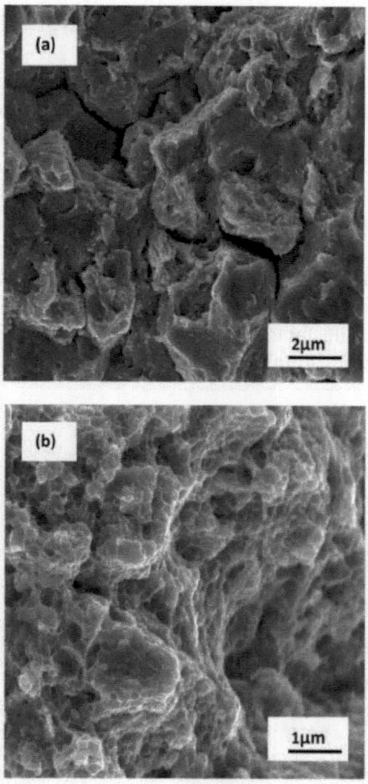

Figure 5. **Scanning electron micrographs of the fatigue fracture surface of alloy steel Tenax™ 310 sample, cyclically deformed at a strain amplitude of 1.00% with a fatigue life of 1,292 cycles, showing**

(a) Region of unstable crack growth showing cracking along the grain boundaries.

(b) Population of dimples covering the overload fracture surface.

4.5.2 Cyclic Strain Amplitude of 0.80 percent [Fatigue life = 6,992 cycles].

At this lower total strain amplitude of 0.80 pct. and resultant fatigue life of 6,992 cycles, overall morphology was normal to the far-field stress axis revealing two distinct regions representing fatigue and overload **(Figure 6-a)**. The transition region (**Figure 6-b**) between fatigue and overload was rough at the microscopic level comprising of a population of fine microscopic cracks reminiscent of brittle mechanisms occurring at the fine microscopic level (**Figure 6-c**). At higher magnification this region of the fracture surface revealed a noticeably array of fine microscopic cracks intermingled with macroscopic cracks (**Figure 6-d).** The region of early microscopic crack growth was noticeably flat and covered with not easily discernible striation-like features coupled with isolated macroscopic cracks. Both these features are clearly indicative of both ductile and brittle mechanisms occurring at the fine microscopic level (**Figure 7-a**). The region of unstable crack growth immediately prior to overload was rough with isolated pockets of striations and featureless transgranular pockets that were intermingled with fine microscopic cracks (**Figure 7-b**).

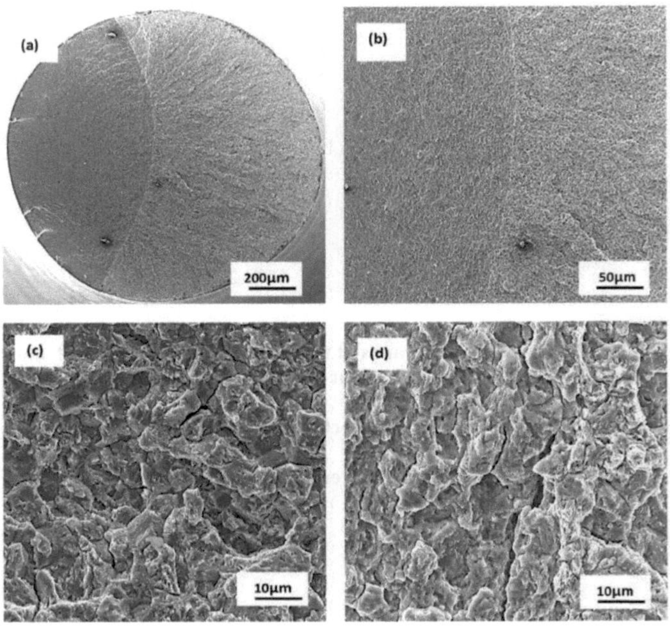

Figure 6. **Scanning electron micrographs of the fatigue fracture surface of alloy steel Tenax™ 310 sample, cyclically deformed at a strain amplitude of 0.80%, with a fatigue life of 6,992 cycles, showing:**

(a) Overall morphology showing distinct regions of fatigue and overload
(b) The transition region microscopically rough
(c) Fine microscopic cracks on the fatigue fracture surface prior to onset of overload
(d) High magnification observation of the region of unstable crack growth reveals both microscopic and macroscopic cracks.

Figure 7. **Scanning electron micrographs of the fatigue fracture surface of alloy steel Tenax^{TM} 310 sample, that was cyclically deformed at strain amplitude of 0.80 pct., with a fatigue life of 6,992 cycles, showing:**
(a) The region of early microscopic crack growth flat and covered with isolated microscopic cracks.
(b) Isolated pockets of striations and transgranular pockets in region of unstable crack growth

4.6 Mechanisms Governing Cyclic Deformation and Fracture Behavior

Under conditions of fully-reversed strain cycling at room temperature (27° C) the mechanisms controlling fracture of this alloy steel can be summarized as follows:

(i) An observable mismatch in strain carrying capability between the soft, ductile and plastically deforming matrix of alloy steel Tenax^{TM}310 and the hard, brittle and essentially elastically deforming microstructural feature, be it the martensite micro-constituent or coarse second-phase carbide particle.

(ii) This causes a large concentration of stress to occur at both the second-phase carbide particle and martensite micro-constituent, thereby enabling in cracking to occur at its interface with the soft metal matrix. The ease of occurrence of fine microscopic cracks suggests that 'local' plastic strain dominates in regions of the hard, brittle and essentially elastically deforming microstructural feature.

Cracking of both the martensite micro-constituent and the coarse second-phase carbide particles occurs normal to the far-field stress axis. The occurrence of plastic strain inhomogeneity at the fine microscopic level (**Figure 8**) within the grains coupled with the presence of coarse second-phase carbide particles is responsible for the traces of intergranular cracking observed.

The fine microscopic mechanisms, which tend to control the variation of cyclic stress amplitude with cycles (N) during fully-reversed strain cycling of this high strength low alloy steel Tenax 310 are dependent on the intrinsic influence of microstructural features and the cyclic strain range. The most likely mechanisms contributing to governing stress response of this high strength alloy steel Tenax ™ 310 can be ascribed to the concurrent and mutually interactive influences of the following events

.

Figure 8: Transmission electron micrographs showing the nature and deformation in test sample that was cyclically deformed at strain amplitude of 1.0 pct. with fatigue life of 1292 cycles.

(a) A gradual increase in the density of dislocations present in the microstructure of this low alloy steel, during fully-reversed cyclic deformation in the plastic range, by the process of dislocation multiplication.

(b) Dislocation-dislocation interactions arising as a direct consequence of an increase in the interaction of a moving dislocation with other dislocations.

(c) An interaction of the moving dislocations with the second-phase constituents i.e., the carbide particles, dispersed through the microstructure coupled with the formation and presence of fine slip bands and their impingement on grain boundaries and resultant stress concentration

The dislocation-dislocation interactions coupled with an interaction of the moving dislocations with the microstructural constituents and grain boundaries is responsible for the hardening or strengthening that occurs at the fine microscopic level during fully reversed cyclic deformation. When the local stress concentration (σ^*) caused by the gradual build-up of dislocations at a second-phase particle exceeds a critical value (σ_C*), microscopic crack initiation is favored to occur through rupture of the hard, brittle and essentially elastically deforming second-phase particle. With continued cyclic straining the fine microscopic cracks that were easily initiated gradually propagate through the microstructure and eventually link to form one or more macroscopic cracks. The early initiation of fine microscopic cracks coupled with their concurrent growth through the microstructure of this alloy steel and culminating in their eventual linkage upon continued cyclic straining results in a noticeable decrease in stress carrying capability of the microstructure, eventually culminating in failure by fracture.

5. Conclusions

Results of an investigation on understanding the influence of strain cycling on mechanical response, deformation and fracture behavior of an alloy steel, i.e., Tenax TM 310, provides the following key findings.

1. Optical microstructure of this alloy steel revealed a combination of carbon-rich, alloy-rich and carbon-depleted regions. The presence of both carbon and alloy content in this steel resulted in the presence of a noticeable volume fraction of second-phase particles and martensite micro-constituent in the alloy-rich and carbon-rich regions.

2. The elastic modulus of this alloy steel is 205 GPa. The yield strength is 1019 MPa and the ultimate tensile strength is 1225 MPa. The higher tensile strength indicates the occurrence of strain hardening beyond yield. The ductility quantified by elongation over 0.5 inch (12.7 mm) gage length was 13.4 pct. The reduction in test specimen cross-section area was 31 pct.

3. This alloy steel exhibited a linear trend for the variation of elastic strain amplitude ($\Delta\varepsilon_e /2$) with reversals-to-failure ($2N_f$), and plastic strain amplitude ($\Delta\varepsilon_p /2$) with reversals-to-failure ($2N_f$), in conformance with the behavior shown by other alloy steels and carbon steels.

4. Fracture morphology was about the same at the macroscopic level over the entire range of cyclic strain amplitudes examined. At the fine microscopic level the alloy steel test specimen that was deformed at the higher cyclic strain amplitude revealed fracture to be mixed-mode, i.e., a combination of ductile and brittle features. At the lower strain amplitude, microscopic observations of the fracture surface revealed predominantly ductile, i.e., a population of fine microscopic voids intermingled with dimples, with only traces of brittle failure mechanisms.

References

1. J.S. Pascover and S.J. Matas: Structure and Properties of Ultra-High Strength Steels, ASTM STP 370, ASTM, Philadelphia, USA, 1965, p. 370.

2. A. J. Baker, E. J. Laura and R. P. Wei in Structure and Properties of Ultrahigh Strength Steels, ASTM STP 370, ASTM,Philadelphia, PA, 1965, p. 3.

3. Strength Materials (edited by V.F. Zackay) John Wiley and Sons, New York, USA, 1965, p. 199,

4. B. R. Banerjee., Structure and Properties of Ultrahigh Strength Steels, ASTM, STP 370, ASTM, Philadelphia, PA, 1965, p. 94.

5. G.E. Pellissier: **Engineering Fracture Mechanics**, 1968, Vol. 1, pp. 55-65.

6. John R. Lowe, Jr.: **Engineering Fracture Mechanics**, 1969, Vol. 1, pp. 55-65.8

7. V.F. Zackay, E.R. Parker, R.D. Goolsby and W.E. Wood: **Nature** Physical Science, 1972, Vol. 236, No. 68, pp. 108-115.

8. D. Huang and G. Thomas: **Metallurgical Transactions**, 1972, Vol. 2, p. 1587

9. R.O. Ritchie, B. Francis and W.L. Server: **Metallurgical Transactions**, 1976, Vol. 7A, pp. 831-841.

10. J.L. Youngblood and M. Raghavan: **Metallurgical Transaction**, Vol. 8A, 1977, pp. 1439-1445.

11. G.Y. Lai, W.E. Wood, R.A. Clark, V.F. Zackay, E.R. Parker, **Metallurgical Transactions**. 5 (1974), 1663–1670.

12. G.Y. Lai, **Materials Science Engineering**, 19 (1975) 153–156.

13 F. Zia Ebrahimi, G. Krauss, **Metallurgical Transactions** 14A (1983) 1109–1119.

14. M. Sudo abd T. Iwai: **International Journal of ISIJ**, Vol. 23, 1983, pp. 284-302.

15 M. Sudio, S. Hashimoto, S. Kambe: **International Journal ISIJ**, Vol. 23, 1983, pp. 303-311.

16 F. Zia Ebrahimi, G. Krauss, **Acta Metallurgica**. 32 (1984) 1767–1777.

17. I. Kim, S. Rachel and W. Dahl: Steel Research, Vol. 58, 1987, pp. 186-190.

18. B.Y. Choi, G. Krauss, D.K. Matlock, Bainite formation and deformation behavior in an intercritically annealed Fe-1.0Mn-0.09C steel, **Scripta Metallurgica**, Vol. 22, 1988, pp. 1575-1580.

19. B.C. Kim, S. Lee, D.Y. Lee, N.J. Kim, **Metallurgical Transactions,** 22A (1991) 1889–1892.

20. D.K. Matlock and G. Krauss, Observation of deformation and transformation behavior of retained austenite in a 0.14C-1.2Si-Mn steel with ferrite-bainite-austenite structure, **Materials Science and Engineering A**, Volume 165, Issue 1, 25 June 1993, pp 1–8

21. Application of Fracture Toughness Parameters to Structural Metals, edited by Herman Greenberg, Gordon and Breach, New York, USA, 1966.

22. R.O. Ritchie, B. Francis, W.L. Server, **Metallurgical Transactions,** 7A (1976) 831–838.

23. W.E. Wood, **Metallurgical Transactions**, 8A (1977) 1195–1199.

24. R.O. Ritchie, R.M. Horn, **Metallurgical Transactions**, 9A (1978), pp. 331–341.

25. B.V. Narasimha Rao, G. Thomas, **Metallurgical Transactions**, 11A (1980) 441–457.

26. D. Thomas, D. Schmatz and W.W. Gerberich: in High Strength Materials (Edited by: V.F. Zackay) John Wiley and Sons, New York, USA, 1965, pp. 199-209

27. R. Hertzberg, Deformation and Fracture Mechanics of Engineering Materials, Second Edition, John Wiley and Sons, 1983,

28. ASTM E-8-06: Tensile Testing of Metallic Materials, American Society of Testing and Materials, Philadelphia, PA, 2006.

29. L. F. Coffin, Jr., and J.F. Travernelli: "The Cyclic Straining and Fatigue of Metals", **Transactions TMS-AIME**, Vol.215 (5), pp. 794-806.

30. S.S. Manson, and M. H. Hirschberg, "Fatigue Behavior of Strain Cycling in the Low and Intermediate Cycle Range," 10th Sagamore Army Materials Research Conference, Raquette Lake, NY, August 13-16, 1963.

31. T. H. Topper, B. I. Sandor, and J. Morrow, "Cumulative Fatigue Damage Under Strain Control" Journal of Materials, Vol. 4, 1969, pp. 189-199.

32. S. S. Manson: NASA Technical Note, National Aeronautics and Space Administration, 1954, pp 2933 - 2940.

33. K. Manigandan, and T. S. Srivatsan: **Journal of Strain Analysis for Design** (in review) 2014.

Acknowledgements

The authors (**TSS and KM**) extend most sincere thanks and overwhelming appreciation to **Carpenter Technology Corporation** (Reading, PA, USA: Program, Manager: ***Mr. Michael Schmidt***, Technical Director: ***Dr. Timothy Anderson***) for providing both material and financial support that made this research endeavor possible.

Fatigue of Materials III: Advances and Emergences in Understanding
Edited by: T.S. Srivatsan, M. Ashraf Imam, and R. Srinivasan
TMS (The Minerals, Metals & Materials Society), 2014

SCANNING ELECTRON MICROSCOPY OF HYDROGEN-EXPOSED 304 STAINLESS STEEL BENDING FATIGUE SAMPLES

M. Phillips[1], K. McLaughlin[1], D. Ojeda[1], E. Luna, M. Burke[1], P. Cruz[1], K. Featherstone[1], E. Petersen[1], P.D. Ferro[1]

Department of Mechanical Engineering
[1] **Gonzaga University**
502 E. Boone Ave., Spokane, Washington (WA), USA

Abstract

Bending fatigue samples were plasma cut and water jet cut from sheets of 304 stainless. Samples were exposed to high pressure hydrogen (138 MPa at 300C for three weeks) or 1 atm hydrogen at room temperature for one week, or were unexposed. Preliminary results from bending fatigue testing show a decrease in fatigue life for hydrogen supersaturated samples. The present investigation shows SEM fractographic imaging from fatigue samples for each of the hydrogen exposure populations. Samples that were not exposed to hydrogen preliminarily appear to show fatigue striations on fractographic images. The ongoing research seeks to discern a difference in failure surfaces between hydrogen-exposed and unexposed samples.

Key Words: Hydrogen embrittlement, bending fatigue, austenitic stainless steel

1. Introduction

As society pushes towards using alternative fuels, hydrogen will be considered as an environmentally-friendly option. Enabling hydrogen-based designs of the future requires continued development of a materials properties database for hydrogen-compatible materials. The current work supports this effort. For example, structures including hydrogen storage tanks and steel pipelines need to be designed to be suitably fatigue resistant. Proper materials selection decisions includes consideration of fatigue resistance following hydrogen exposure.

Two possible mechanisms that are proposed for explaining hydrogen embrittlement are hydrogen-enhanced localized plasticity (HELP) and the lattice decohesion effect. The research, under the auspices of the hydrogen embrittlement project at Gonzaga University, seeks to investigate these theories through image analysis of fracture surfaces obtained using scanning electron microscopy. Samples of 304 stainless steel samples that have been failed under bending fatigue testing have been preliminarily examined using SEM. The samples have been variously exposed to differing amounts of hydrogen, including: no hydrogen exposure (control), 1 week hydrogen exposure at 1 atm in the Gonzaga University Hydrogen Lab, and supersaturated exposure at Sandia National Laboratories. The research was focused on determining if higher levels of hydrogen exposure correlated with visually-detectable high levels of localized microscopic plasticity. The research also attempted to look for a possible correlation between the presence of fatigue striations and the amount of hydrogen exposure. Analysis of this data may give insights on how hydrogen affects crack growth rate and degrades mechanical properties of the material.

An ongoing research focus of the Hydrogen Lab at Gonzaga has been in bending fatigue. Bending fatigue samples from previous investigations [1-3] have used samples that were plasma cut from 0.9mm sheet stock. Following plasma cutting, samples had been polished to remove slag from the cutting process. Samples were generally cut in the same orientation from a given sheet of rolled sheet stock. The actual orientation of the sheet (parallel or perpendicular to rolling direction) was previously not studied.

The current work has initiated an effort to understand the possible rolling direction effect. Samples were cut from sheet stock in each of two orthogonal directions. Other changes to the sample generation process included waterjet cutting and a slight change to the radius at the specimen grip.

2. Procedure

Two types of bending fatigue samples were tested. Each were cut from 0.9 mm thick sheets of type 304 stainless steel. One style was plasma cut at the machine shop at Gonzaga University, such that all specimens were oriented in the same direction relative to the rolling direction of the stainless sheet. The other style was waterjet cut at Northwest EDM (Spokane Valley WA), such that specimens were cut from each of two orthogonal directions.

The plasma-cut specimens were belt-ground along the periphery to remove slag. Following belt grinding, the plasma cut specimens were progressively hand sanded down to 600 grit sand paper. The waterjet specimens were lightly sanded with 600 grit sand paper. Hydrogen exposure conditions for each type of bending fatigue specimen included no exposure, one week at 1 atm hydrogen and 138 MPa at temperatures as high as 375°C for 15 days ('supersaturation conditions'). The typical hydrogen concentration in 300 series stainless steels after similar conditions is approximately 140 ppm, by weight [5]. Other hydrogen charging protocols are described by Mine et al. [6] The bending fatigue experimental equipment and procedures may be found in previous papers [1-3].

Figure 1 shows a photo of both types of specimens. The plasma cut specimen is labeled as specimen 1 in the photo, and the waterjet cut specimen as specimen 2. The difference between the two specimens is in the transition from the specimen grip to the bending portion.

The bending fatigue tests were performed using a VSS-40H bending fatigue testing machine (Fatigue Dynamics, Walled Lake MI). The cycling frequency was between 300 and 350 cycles per minute. After the specimen failed, it was removed and the location of the failure was identified by measuring the distance from the start of the specimen radius to the middle of the failure location on specimen. The fatigue ratio of the bending was R = -1.

To prepare samples for SEM, the fatigue specimens were had been failed in the bending fatigue testing machine such that the crack had propagated approximately 75% through the specimen. The failed specimens were prepared by bending them to allow for mirror-image failed surfaces to be adjacent.

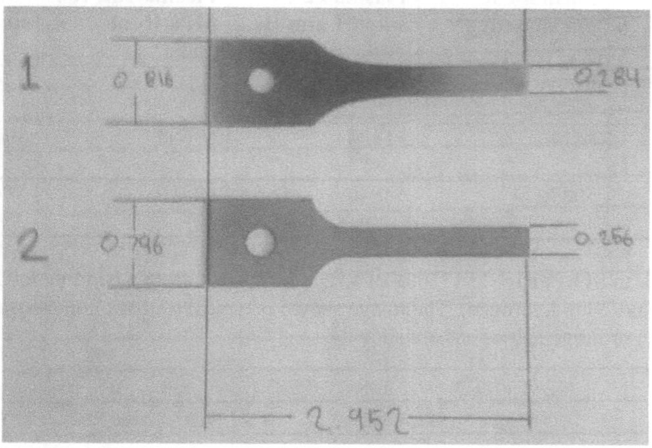

Figure 1. Photo of bending fatigue specimens. The top specimen is plasma cut at Gonzaga from 0.9 mm sheets of 304 stainless. The lower specimen is waterjet cut from the same material. The dimensions shown are in inches.

3. Results

Figure 2. shows the bending fatigue results for specimens that were tested at a nominal maximum bending stress amplitude of 220 MPa. The results compare the differences between the older style plasma cut specimens and the newer design waterjet cut specimens. The effect of high pressure hydrogen exposure can be seen by comparing the plasma cut specimen populations (the first three shown) to each other. The data for the 138 MPa hydrogen exposure (circles) has a lower average cycles to failure than that for the other two plasma-cut specimens' populations (diamonds and triangles).

Table I shows the statistical summary for the data shown in figure 2. The standard deviation for the unexposed waterjet cut specimens is much lower than that for the other populations. Also, the difference between the unexposed plasma-cut specimens and the plasma-cut specimens that were exposed to 1 atm hydrogen are similar.

Figure 3 compares the waterjet specimens for each of the two different orthogonal directions from which they were cut from the original sheet. The nominal bending stress for the data shown is 220 MPa. The data does not appear to show that the orientation affects the bending fatigue behavior, for samples that were not exposed to hydrogen.

Table I. Fatigue summary results at 220 MPa

	Plasma-cut, unexposed	**Plasma-cut, 1 wk at 1 atm H_2,**	**Plasma-cut, 138 MPa H_2 at 300°C**	**Waterjet cut, unexposed**
count	5	10	7	17
average	84140	91470	53600	14105
standard deviation	50890	54840	62810	1060

Figure 4 shows an SEM image at 13100x of a fracture surface from a bending fatigue specimen that was not exposed to hydrogen. The image shown is typical of that of unexposed samples. A series of ridges, or fatigue striations are shown

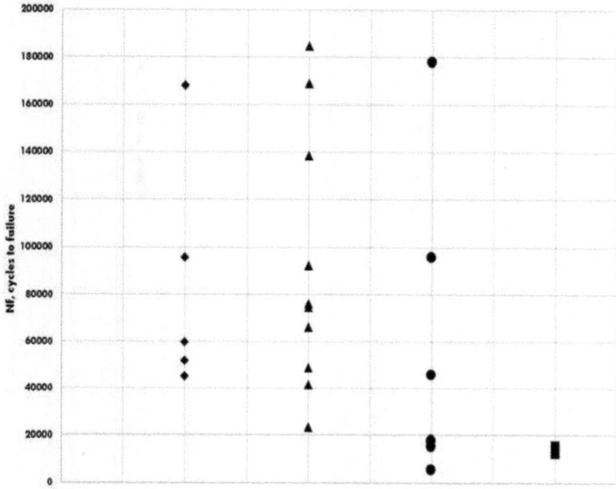

Figure 2. Cycles to failure (N_f) for bending fatigue specimens tested at a nominal maximum bending stress of 220 MPa. The different populations represented are for plasma cut specimens with no exposure to hydrogen (diamonds), plasma cut specimens exposed to 1 atm hydrogen for 1 week (triangles), plasma cut specimens that were exposed to 138 MPa hydrogen at 300°C (circles) and waterjet cut specimens that were not exposed to hydrogen (squares). The waterjet cut specimens have a much lower cycles to failure and are more predictable, possibly as a function of the specimen geometry.

Figure 3. Bending fatigue data for waterjet cut specimens from each the two directions from which the specimens were cut. The populations represent cycles to failure for samples that were cut from the sheet in one direction (circles) and that for samples cut from a direction 90° from the first direction (triangles). The maximum nominal stress during bending for the data shown is 220 MPa. The samples were not exposed to hydrogen.

Figure 4. SEM image of a failure surface for a bending fatigue specimen that was not exposed to hydrogen. Fatigue striations may be observed traveling from the bottom right to the top left of the image. The striations appear to be fairly continuous and relatively closely spaced.

Figure 5 shows an SEM image of a failure surface for a bending fatigue specimen that was exposed to 138 MPa hydrogen. The image shown is representative of failed specimens that were exposed to high pressure hydrogen. Fatigue striations are observable, although less likely to be on continuously planar surfaces.

Figure 5. Fatigue crack surface of a specimen subjected to high-pressure hydrogen exposure. Fatigue striations are much less identifiable in this specimen, though some can be seen on various surfaces. This result is fairly typical for specimens exposed to high-pressure hydrogen

Comparing the representative SEM images shown in Figure 4 and Figure 5 allows for the preliminary observation that fatigue striations are less prominent on samples that have been subjected to high-pressure hydrogen exposure compared to specimens that were not exposed to hydrogen. This result may preliminary suggest that hydrogen affects fatigue-crack growth rate as a function of how fatigue striations relate to crack growth rate.

4. Conclusions

High pressure hydrogen appears to reduce the bending fatigue life of 304 stainless steel. The affect of 1 atm hydrogen for one week does not appear to have a noticeable affect on the bending fatigue life of 304 stainless by comparing the bending fatigue life for unexposed samples. Samples that were waterjet cut from 304 stainless steel sheet in two mutually orthogonal directions do not appear to have a difference in bending fatigue life. Preliminary SEM results show that images from specimens that were exposed to high pressure hydrogen appear to different than that of unexposed specimens. The largest difference is seen by the morphology of the fatigue striations.

Acknowledgments

The authors acknowledge the support provided by Gonzaga University, and by the faculty, staff and students of the School of Engineering and Applied Sciences (SEAS). The authors particularly acknowledge the help of Floyd Grillo, Trevor Longbottom and and others in the shop for helping generate test samples.

References

1. P. Ferro, "Effect of Hydrogen on Bending Fatigue Life for Materials Used in Hydrogen Containment Systems", Advances in Materials Science for Environmental and Nuclear Technology II, ed. S.K. Sundaram, T. Ohji, K. Fox, E. Hoffman, *Ceramic Transactions*, v. 227, pp. 39-49, American Ceramic Society Publication, Wiley ISBN 978-1-118-06000-1 (2011).

2. P. Ferro et al., "Fatigue Testing of Hydrogen-exposed Austenitic Stainless Steel", Advances in Materials Science for Environmental and Energy Technologies, *Ceramic Volumes* (2012).

3. P. Ferro, R. Miresmaeili, R. Mitra, J. Ross, W. Tiedemann, C. Hebert, D. Howard, "Hydrogen-exposed Austenitic Stainless Steel Welded Specimens in Bending and Rotational Bending Fatigue", MS&T 2012 Conference, Pittsburgh PA, October 2012, *Ceramic Volumes* (2013).

4. H.G. Nelson, "Hydrogen Embrittlement", Treatise on Materials Science and Technology, v. 25, Academic Press, ISBN 0-12-341825-9 (1983).

5. C. San Marchi, B.P. Somerday, X. Tang, G.H. Schiroky, Effects of Alloy Composition and Strain Hardening on Tensile Fracture of Hydrogen-precharged Type 316 Stainless Steels, **Int'l J. Hydrogen Energy**, **33,** 889-904, (2008).

6. Y. Mine, K. Tachibana, Z. Horita, Effect of High-Pressure Torsion Processing and Annealing on Hydrogen Embrittlement of Type 304 Metastable Austenitic Stainless Steel, **Met. and Mat. Trans. A, 41A (**2010).

Fatigue of Materials III: Advances and Emergences in Understanding
Edited by: T.S. Srivatsan, M. Ashraf Imam, and R. Srinivasan
TMS (The Minerals, Metals & Materials Society), 2014

INFLUENCE OF LONG TERM AGEING ON FATIGUE CRACK GROWTH BEHAVIOR OF P91 STEEL AT DIFFERENT TEMPERATURES

M. Nani Babu, S. K. Swain, B. Shashank Dutt, S. Venugopal, G. Sasikala*

Metallurgy & Materials Group
Indira Gandhi Centre for Atomic Research,
Kalpakkam, 603102, **India**
*gsasi@igcar.gov.in

Abstract

Fatigue crack growth behavior (FCG) of Modified 9Cr-1Mo (P91) steel has been evaluated at 300, 653 and 823 K in normalized and tempered (unaged) and in subsequently aged at 823 K for 20000 h (aged) condition in both Paris and threshold regimes. The effects of test temperature and ageing on the FCG resistance of the material have been rationalized by examining the behavior using different measures of driving forces, namely, the applied and effective stress intensity factor ranges ΔK and ΔK_{eff}, and the crack tip stress intensity factor range $\Delta K_{tip,eff}$ that takes in to account the stress shielding effects also. The variations in the extent of crack closure in the threshold regimes and the crack path tortuousness at different temperatures indicated that in addition to plasticity, roughness induces closure at 300 and 653 K, while at 823 K possibly oxide induced closure too takes place. Crack growth rates for the aged material are higher than that of unaged P91 steel for all the temperatures examined, irrespective of the choice of driving force and is attributed to the loss in strength as well as formation of embrittling Laves phase.

1. Introduction

Modified 9Cr-1Mo steel (P91) has been chosen for the steam generator of the sodium cooled prototype fast breeder reactor (PFBR) which is under construction at Kalpakkam, India, in view of its excellent combination of creep and stress corrosion cracking resistance, thermal conductivity, and low thermal expansion. The normal operating temperatures seen by the steam generator (SG) components of PFBR are in the range of 653 to 823 K. The high temperature tensile, creep, and toughness properties and ambient temperature fatigue crack growth (FCG) properties influenced by thermal ageing are available for plain 9Cr–1Mo (P9) steel [1-4]. The low cycle fatigue (LCF), quasi static and dynamic fracture behavior have been extensively studied [5-7]. However, studies on the FCG behavior of P91 steel at the service temperatures are limited. During operation, the steam generator components are subjected to flow-induced vibrations and thermal oscillations that cause fatigue loading. In the steam generator, the secondary sodium in temperature range of 628-798 K (in the shell side) and water in temperature range of 508-763 K (in the tubes) are separated by thin walls of the tubes. Any through thickness crack in the tubes or shell would lead to catastrophic failure of steam generator components due to the highly exothermic nature of the reaction between sodium and water/moisture in the air. Thus, integrity of these components is of utmost importance to the safety of SG. For assessing the integrity of the SG components the fatigue crack growth properties are necessary. Therefore, a campaign was initiated to study the fatigue crack growth behavior of indigenously developed P91 steel at different temperatures, viz., 300, 653 and 823 K.

The driving force for FCG has been considered to be the stress intensity factor (SIF) range $\Delta K = K_{\max} - K_{\min}$, the difference of maximum and minimum SIFs corresponding to the peak and trough of the load cycle. For intermediate crack growth rates, a power law relation between the FCG rate (da/dN) and ΔK was proposed by Paris et al [8] in the early studies on FCG, which became very popular and known as Paris equation:

$$da/dN = C\Delta K^m \qquad (1)$$

where the coefficient, C, and the exponent, n, are material constants. This regime, i.e., the Paris regime, is generally considered to be insensitive to microstructure, environment and load ratio ($R = K_{\min}/K_{\max}$). Later, several versions of the Paris equation were used to describe the FCG behavior of various materials in different test and metallurgical conditions, by modifying the driving force parameter to account for phenomena like crack closure [9] and crack tip shielding [10] etc. As ΔK is gradually reduced, the crack growth rate drastically decreases, approaching a threshold value ΔK_{th} at which da/dN approaches zero [11]. Microstructure, environment, temperature, R-ratio, crack tip constraints, crack closure, overloads, cold work etc are found to influence ΔK_{th}; in addition, ΔK_{th} is sensitive to the method of measurement, and specimen geometry [9,10,12,13].

In the present investigation, the influence of ageing at elevated temperatures on FCG behavior of the P91 steel in Paris and threshold regimes in the temperature range of 300 - 823 K is considered. The FCG mechanisms as well as crack closure and crack tip shielding effects have been examined for unaged and aged material.

2. Experimental

2.1 Materials and Specimen Preparation

The P91 steel was received in the form of 24 mm thick plate in normalized (1333 K/1h) and tempered (1053 K/1h) condition. Two sets of specimen blank were cut from this plate of which one set was subjected to ageing at 823 K for 20000 h in resistance heated furnaces with a temperature control of ±3 K. The chemical composition is given in Table 1. The microstructural changes on ageing not resolvable by optical microscopy. The typical tempered martensitic microstructure of normalized and tempered P91 steel is shown in Fig. 1(a). Compact Tension (CT) specimens of 20 mm thickness were fabricated from the P91 steel blanks in both unaged and aged conditions with the notch made using electric discharge machine (EDM). A schematic view of the CT specimen is shown in Fig. 1(b). The specimens were subjected to fatigue pre-cracking to obtain a sharp and reproducible crack.

Table 1. Chemical composition of the P91 steel

Element	C	Mn	Si	P	S	Cr	Ni	Mo	V	Nb	Al	N	Fe
Wt%	0.097	0.37	0.31	0.018	0.0047	9.29	0.38	0.92	0.26	0.08	0.006	0.057	Bal

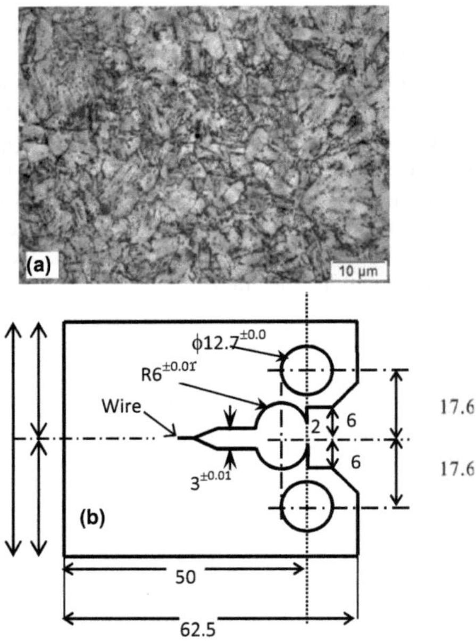

Figure 1. (a) Microstructure of P91 steel (b) CT Sample

2.2 Fatigue Crack Growth Test

The FCG tests on CT specimens were conducted at $R = 0.1$ and a frequency of 15 Hz employing the ASTM procedure [11] at different temperatures with a fully automated servo hydraulic machine. For high temperature tests, the specimens were enclosed in a resistance-heated furnace with a temperature control of better than ±2 K. The crack growth is measured online using a suitably calibrated direct current potential drop (DCPD) system. The opening and closing loads have been estimated from the load (P) - load line displacement (δ) data as per the ASTM standard as detailed in [14].

3. Results

3.1 Fatigue crack growth curves

FCG plots for all the three temperatures without and with closure correction are presented in Fig. 2 (a) & (b) respectively for the unaged and aged P91 steel. Significant differences between unaged and aged materials are observed in Paris and threshold regimes presented in these plots. Clearly, for the same levels of applied □*K* ranges, the FCG rates for the aged material are higher at all the temperatures. The threshold levels of ΔK too are significantly lower for the aged material. Further, for both the unaged and aged materials, the FCG rates for the intermediate temperature of 653 K are lower than that for 300 K at the higher end of □*K* levels, whereas those for 823 K are an order higher.

The FCG results incorporating crack closure effects are presented in Fig. 3 (a) and (b). The data clearly indicate the degradation in FCG resistance with ageing. There is another notable difference between the unaged and aged materials. The trends with respect to temperatures remain the same for the unaged material even after incorporating the closure correction, while for the aged material, the FCG rates are seen to increase with temperature monotonically. The extent of closure at 653 K is significantly higher in the aged material. The variations in the FCG parameters are discussed in detail in the following section.

3.2 Fatigue crack growth parameters in Paris and threshold regimes

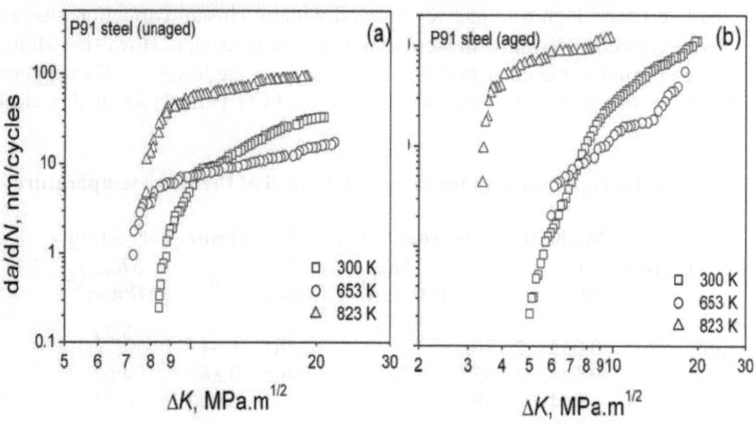

Figure 2. **Fatigue crack growth behavior of (a) unaged and (b) aged P91 steel without crack closure correction at different test temperatures**

The values of Paris constants C and n obtained using ΔK and ΔK_{eff} to correlate

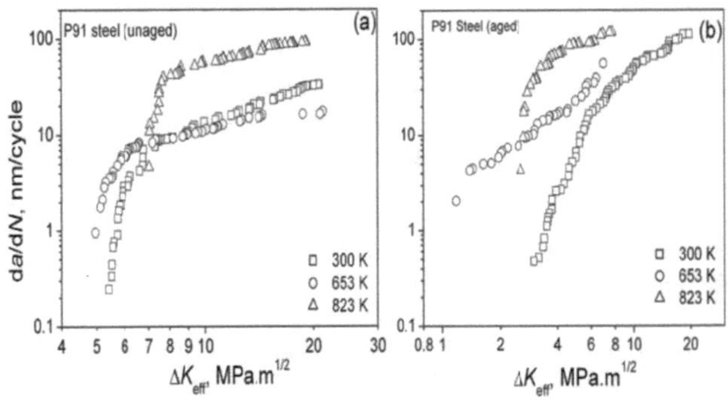

Figure 3 Fatigue crack growth behavior of (a) unaged and (b) aged P91 steel with crack closure correction at different test temperatures

the FCG rates for the three temperatures are presented in Tables 2 and 3 for unaged and aged material respectively. For the unaged material, irrespective of whether ΔK or ΔK_{eff} is used for correlating FCG rates, it is observed that there is a monotonic increase in C with temperature. On the other hand, value of n is more or less similar at 653 K and 823 K. For the aged material, the low C and high n at 653 K observed without closure correction changes on incorporating closure correction resulting in increasing C and decreasing n. However, while ΔK_{th} is higher at 653 K compared to the other temperatures, $\Delta K_{th,eff}$ is the lowest at this temperature. Thus, the contributions of ageing and crack closure on the FCG parameters of this steel are complex.

Table 2. The FCG parameters for unaged P91 steel at the three temperatures

	Without closure correction			With closure correction		
Temperature(K)	**C (nm)**	**n**	**ΔK_{th} (MPa.m$^{1/2}$)**	**C (nm)**	**n**	**$\Delta K_{th,eff}$ (MPa.m$^{1/2}$)**
300	0.053	2.156	8.28	0.581	1.347	5.27
653	0.566	1.057	7.35	1.468	0.889	4.62
823	6.164	0.939	7.46	6.841	0.900	6.35

Table 3. The FCG parameters for aged P91 steel at the three temperatures

Temperature (K)	Without closure correction			With closure correction		
	C (nm)	n	ΔK_{th} (MPa.m$^{1/2}$)	C (nm)	n	$\Delta K_{th,eff}$ (MPa.m$^{1/2}$)
300	0.310	1.966	4.70	1.947	1.376	2.75
653	0.035	2.485	5.42	3.448	1.37	0.84
823	13.440	0.964	3.13	18.476	0.923	2.17

4. Discussion

4.1 Crack Closure Mechanisms

Various crack closure mechanisms such as plasticity induced closure (PIC), roughness induced closure (RIC), oxide-induced closure (OIC) etc, have been discussed in the literature [12], which are known to be more significant in the near threshold regime. All these can directly/indirectly influence the FCG behavior depending on temperature. Extensive oxidation occurs at a temperature 823 K which gives rise to significant OIC effects, increasing the extrinsic FCG resistance of the material. PIC, as a first approximation, is proportional to the plastic zone size [2] which in turn depends inversely on the yield stress, $r_p = (1/2\pi)(K_{max}/\sigma_y)^2$, for plane stress condition and $r_p = (1/6\pi)(K_{max}/\sigma_y)^2$ for plane strain condition. Therefore, the PIC is expected to increase with increasing temperature (decreasing σ_y). The extent of closure in the present study, decreased with increase in test temperature for the unaged material; 3.01, 2.73 and 1.11 MPa.m$^{1/2}$ at 300, 653 and 823 K respectively, are estimated (ΔK_{th} -$\Delta K_{th,eff}$) from the data presented in Table 2. The estimated plastic zone sizes corresponding to the respective ΔK_{th} were similar for 300 and 653 K (10.7 and 10.6 μm), whereas that at 823 K was nearly twice (18.3 μm). Therefore, it is clear that other factors too contribute to closure. This decrease in the extent of closure with increasing temperature is due to the role of RIC as highlighted in earlier study [14,15]. Highly tortuous crack paths (high roughness of fracture surface) were observed at 300 and 653 K, whereas quite straight crack path was observed at 823 K. Change in slip character from planar to wavy with increasing temperature was considered to be the reason for this.

Now, in the case of aged material (Table 3), the closure extents in the threshold regimes are 1.95, 4.58 and 0.96 respectively at 300, 653 and 823 K. The plastic zone sizes estimated corresponding to ΔK_{th} are 4.7, 9.0 and 4.1 μm respectively. Very high levels of closure at 653 K may be partially due to this. Further, at 653 K, a highly intergranular crack path with multiple branches was observed as compared to lower levels of branching and tortuousness at 300 K, and nearly straight crack path at 823 K (Fig. 4). The explanation with respect to change in slip character with temperature is applicable here too. Planar slip leads to a highly serrated and tortuous crack path as shown in Fig. 4 (a) & (b) resulting in significant roughness induced crack closure, while wavy and homogeneous slip leads to a rather straight crack path and smooth fracture surface in Fig. 4 (c) and relatively less roughness-induced crack closure is evidenced.

4.2 Crack Tip Shielding

The discussion so far considers ΔK_{eff} as the driving force for fatigue crack growth. In certain materials like ceramics, intermetallics, and composites, stress shielding at the crack tip by various phenomena such as bridging, interlocking and branching of the cracks has been observed to lead to improved crack growth resistance [13]. In such cases, the crack tip stress intensity factor range (ΔK_{tip}) rather than ΔK_{eff} should be considered as the appropriate driving

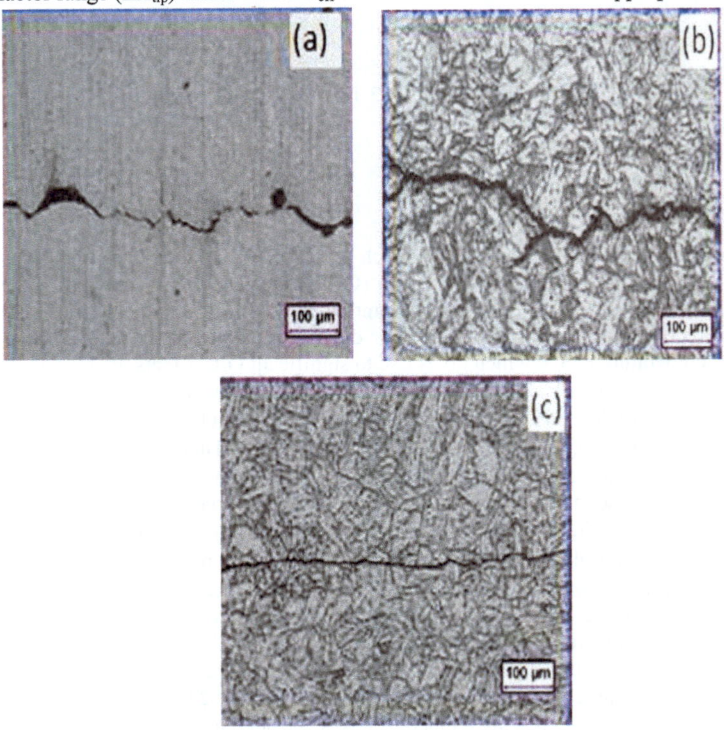

Figure **4.** **Optical micrographs showing decreasing crack path tortuousness with test temperature (a) 300 K (b) 653 K (c) 823 K of aged P91 steel**

force for crack growth. In earlier studies, the present authors have examined the fatigue crack growth behavior of different steels including P91 [14,16,17] as a function of effective crack tip SIF in line with procedure proposed by Mutoh [18]. Interlocking and branching of the cracks have been observed at 300 and 653 K (Fig. 4) in the present study too. Therefore, the possible effects of crack tip shielding on the observed FCG behavior has been examined. The relevant equations and the detailed procedure to estimate are presented in [17].

The stress intensity factor seen by the crack tip K_{tip} is experimentally evaluated by measuring the load line displacement δ. A linear relationship exists between $K_{\max}$ and δ_{ideal}, the load line displacement expected for ideal cracks without stress shielding. Assuming that the same proportionality exists between K_{tip} and δ_{exp}, the δ measured during fatigue crack growth tests, which includes the stress shielding effect,

$$K_{tip} = \frac{\delta_{\exp}}{\delta_{ideal}} K_{\max} \qquad (2)$$

Thus,

$$\Delta K_{tip,eff} = K_{tip} - K_{cl} \qquad (3)$$

where K_{cl} is the crack closure stress intensity factor. The relationship between $\Delta K_{tip,eff}$ and da/dN is presented in Fig. 5. At all the test temperatures, a Paris type relationship is observed for almost the entire range of $\Delta K_{tip,eff}$. The values of FCG parameters estimated from this plot are presented in Table 4 for both the materials. Further, it is seen that the "anomaly", viz., lower crack growth rate at 653 K compared to 300 K in the Paris regime persists for both the materials. Also, the curves for different temperatures appear to intersect at approximately the same value of ΔK (~ 1.5 MPa.m$^{1/2}$). The temperature dependence of crack growth rate vs the driving force, i.e., $\Delta K_{tip,eff}$, also has decreased with ageing. The reasons/implications of these observations need to be examined.

Table 4. FCG parameters with shielding correction

Temperature (K)	Unaged C (nm)	Unaged n	Unaged $\Delta K_{tip,eff,th}$ (MPa.m$^{1/2}$)	Aged C (nm)	Aged n	Aged $\Delta K_{tip,eff,th}$ (MPa.m$^{1/2}$)
300	6.689	0.760	0.500	9.818	1.067	0.375
653	7.626	0.384	0.365	2.124	1.490	0.573
823	33.202	0.514	0.867	16.009	1.100	1.486

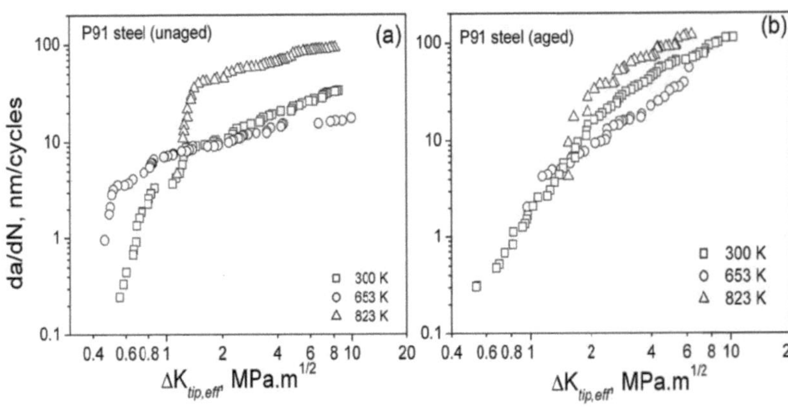

Figure 5. Fatigue crack growth behavior of (a) unaged and (b) aged P91 steel at different test temperatures with crack tip shielding

4.3 FCG Parameters

The inverse linear relation between log(C) and n and the reasons thereof have been discussed by many researchers [19-22]. A plot of log(C) vs n presented in Fig. 6 which brings out the effect of ageing on the FCG parameters. All three sets of data ((i) with out any correction, (ii) with closure correction and (iii) with shielding correction) are included. The data for aged material is clearly above that for the unaged material. Further, the linear relation between log(C) and n is clear for the aged material for all three sets of data. On the other hand, for the unaged material there is a scatter with one point deviating from the trend in each set of data. This point belongs to the intermediate temperature of 653 K. The data for unaged material excluding this point can be seen lie in a line parallel to that for the aged material. The fact that the data for 653 K for the unaged material shows a deviation from the general trend may be attributed to the possible effects of dynamic strain ageing (DSA) that is known to be dominant in P91 at this temperature [23]. However, this point needs to be examined further.

4.4 Effects of ageing

It is seen that ageing at 823 K leads to a decrease in the FCG resistance of the P91 steel. This may be attributed to several factors. Significant microstructural changes take place in this material during long term ageing at high temperatures. The most important is the coarsening of the carbides and precipitation of other phases such as Laves phase etc. [24,25]. This can reduce the solid solution strengthening effect by reducing the content of the solid solution strengtheners in the solution. Also, the effectiveness of the precipitates in strengthening is reduced due to coarsening. Further, the secondary phases, e.g., Laves that form along the boundaries are embrittling in nature. Extensive grain boundary cracking observed in the aged material corroborates this. Also, sub-microstructural changes such as recovery of the lathe structure can take place due to thermal ageing. All these factors can lead to a reduction in FCG resistance. Since ageing leads to removal of the elements responsible for DSA from solid solution in the form of precipitates, the DSA effect is expected to be insignificant in aged material.

5. Summary and Conclusions

FCG behavior of P91 steel has been evaluated at ambient and elevated temperatures in

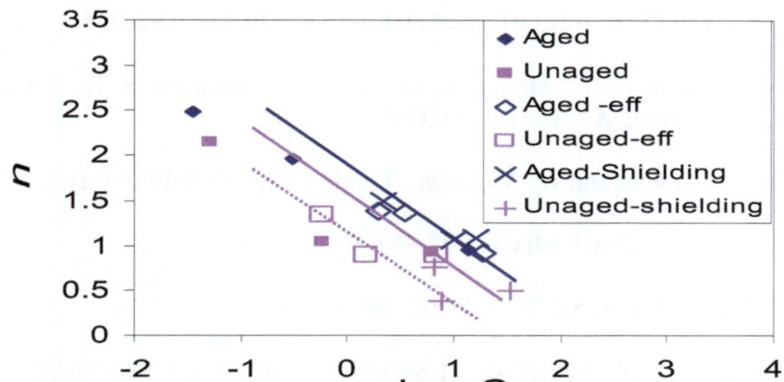

Figure 6. Paris constant (log(C)) as a function of exponent (*n*)

normalized and tempered condition and after subsequent ageing at 823 K for 20000 h. The crack growth rate for aged material is found to be higher than that of unaged at all the test temperatures. Crack closure mechanisms in these conditions have been examined. Roughness induced crack closure is dominant at 300 and 653 K, while at 823K it is not significant due to straight crack path which has been attributed to wavy and homogenous slip character. The effective SIFs at crack tip have been evaluated taking in to account the shielding effects and used to correlate the crack growth rate data. Even with effective crack tip SIF ($\Delta K_{tip,eff}$), the deviation of 653 K data for unaged material from linear log(C) vs. n relation remains, indicating possible DSA effects at this temperature. Absence of any such deviation in the data for the aged material is in agreement with this. Ageing induced changes in microstructure, especially the precipitation of Laves phase and reduced strength due to loss in precipitation and solid solution strengthening appear to be the reasons for degradation in FCG resistance due to ageing.

References

1. Kamel Makhlouf and J. W. Jones, **Intl. J. Fatigue**, 14, 97-104 (1992).

2. Vibhor Chaswal, G. Sasikala, S. K. Ray, S. L. Mannan and Baldev Raj, **Mat. Sci. Engg**. A 395, 251-64 (2005).

3. B. A. Senior, F.W. Noble and B. L. Eyre, **Acta Metall.**, 36, 1855-62 (1988).

4. W. H. Lee, R. K. Shiue and C. Chen, **Mater. Sci. Engg** A, 356, 153-61 (2003).

5. Vani Shankar, M. Valsan, K. Bhanu Sankara Rao, R. Kannan, S. L. Mannan and S. D. Pathak, Mater. Sci. Engg. A, 437, 413-22 (2006).

6. G. Sasikala and S. K. Ray, **Mater. Sci. Engg. A,** 479, 105-11 (2008).

7. S. Sathyanarayanan, A. Moitra, K. G. Samuel, G. Sasikala, S. K. Ray and V. Singh, **Mater. Sci. Engg. A** 488, 519-28 (2008).

8. P. Paris, M. Gomez and W. Anderson, **Trend. Engg**. 13, 9-14 (1961).

9. W. Elber, **Engg. Fract. Mech**. 2, 37-45 (1970).

10. R. O. Ritchie, **Mater. Sci. Eng**. A 103, 15-28 (1988).

11. E647-08e1, Annual Book of ASTM Standards, ASTM, PA, USA (2010).

12. S. Suresh, Fatigue of Materials, Cambridge University Press, (1991).

13. R. O. Ritchie, C. J. Gilbert and J. M. McNaney, **Intl. J. Solids and Stru**. 37, 311-29 (2000).

14. M. Nani Babu and G. Sasikala, **Materials Performance and Characterization**, Vol. 3, No. 2, 182-209 (2014) doi:10.1520/MPC20130055. ISSN 2165-3992.

15. M. Nani Babu, B. Shashank Dutt, S. Venugopal, G. Sasikala, A. K. Bhaduri, T. Jayakumar, Fatigue crack growth behaviour of P91 steel in the temperature range of 300-

823 K, p. 295, in : C. Berger, H.-J. Christ (Eds.), Fifth Int. Conf Very high cycle fatigue (VHCF5), DVM, Berlin, (2011).

16. M. Nani Babu, B. Shashank Dutt, S. Venugopal, G. Sasikala, A.K. Bhaduri, T. Jayakumar and Baldev Raj, **Mater. Sci. Eng. A** 527, 5122–29 (2010).

17. M. Nani Babu, G. Sasikala, B Shashank Dutt, S Venugopal, A. K. Bhaduri, T. Jayakumar, **Nucl. Eng. Design** 269, 103-7 (2014).

18. Y. Mutoh, Akhmad A Korda, Y. Miyashita and T. Sadasue, **Mater. Sci and Eng**. A 468–470, 114-9 (2007).

19. K. Tanaka, C. Masuda and S. Nishijima**, Scripta metallurgica** 15, 259-64 (1981).

20. F. Bergner and G. Zouhar, **Int. J. Fatigue** 22, 229-39 (2000).

21. Johannes Weertman, **Mater. Sci. Eng**. A 468-70, 59-63 (2007).

22. Alberto Carpinteri, Marco Paggi, **Eng. Frac. Mech**. 74, 1041-53 (2007)

23. M. Nani Babu, G. Sasikala, B. Shashank Dutt, S. Venugopal, S. K. Albert, A. K. Bhaduri and T. Jayakumar, **Int. J. Fatigue** 43, 242-45 (2012).

24. V. Thomas Paul, S. Saroja and M. Vijayalakshmi, **J. Nucl. Mater**. 378, 273-81 (2008).

25. S. Sathyanarayanan, Joysurya Basu, A. Moitra, G. Sasikala and V. Singh, **Metall. Trans**. 44A, 2141-55 (2013).

Fatigue of Materials III: Advances and Emergences in Understanding
Edited by: T.S. Srivatsan, M. Ashraf Imam, and R. Srinivasan
TMS (The Minerals, Metals & Materials Society), 2014

MULTISTAGE STRENGTH DEGRADATION IN S25C STEEL UNDER TORSIONAL CYCLIC LOADING AND ITS ENGINEERING APPLICATIONS

Zihai Shi[1], Masaaki Nakano[1], Cuiping Liu[1]

[1]Research and Development Centre
Nippon Koei Co., Ltd.
2304 Inarihara, Tsukuba
Ibaraki 300-1259, **Japan**
Email: a4739@n-koei.co.jp (SHI)

Abstract

An experimental investigation on the fatigue mechanism of metal is reported. In this experimental study, multiple identical test specimens were first preloaded with torsional cyclic loads of increasing number of load cycles, and then tension tests were carried out to obtain their tensile strengths. The obtained test results presented strong experimental evidence for the existence of multistage strength degradation processes in metals under cyclic loading, and the strength degradation was clearly triggered by an abrupt change of cracking behaviour. To improve the theoretical basis of fatigue design by reducing data dispersion in fatigue tests, a modified *S-N* curve for the region near the fatigue limit was proposed. The grade-D-based section of the curve represents the number of load cycles required for the strength degradation to reach grade D. A strength degradation diagram based on the level of stress and the range of load cycles was also proposed for application.

1. Introduction

Fatigue, as a fundamental engineering problem, has been the focus of extensive research since the mid nineteenth century. It is one of the most challenging issues in modern science, combining extreme academic difficulties with enormous engineering consequences. As shown by the few influential references in [1-13] from the vast body of research on the fatigue and fracture of engineering materials, fatigue research is based on two schools of theory: the *S-N* curve and fracture mechanics. While fatigue theories based on the *S-N* curve represent a phenomenological approach and a simple design philosophy that failure can be avoided if this curve is observed as a safety limit in design, the fracture mechanics approach studies the fracture processes of materials under fatigue and thus has the potential to clarify the fundamental mechanism of fatigue that may lead to better design practice.

The multistage strength degradation theory, which has recently emerged from fracture mechanics studies on the material and structural behaviour of concrete, provides a clear description of the fatigue mechanism [14, 15]. According to this theory, fatigue is caused by the sporadic sudden change of cracking behaviour in a system under cyclic loading, leading to intermittent strength reduction of the system and its eventual failure. As metal is the main engineering material suffering most from fatigue failure, this study aims to experimentally verify the new theory on the fatigue mechanism of metal.

First, the scheme of the experimental study is explained. Then, test results on multistage strength degradation and strain energy variation are presented, using the severity of cracking of test specimens as an index for damage classification. The test results present strong experimental evidence for the existence of multistage strength degradation processes in metals under cyclic loading, and the strength degradation is caused by the change of cracking behaviour. To improve the theoretical basis of fatigue design, a partially-modified *S-N* curve based on the concept of strength degradation is proposed. Further, to facilitate maintenance work on existing structures and machine parts, a strength degradation diagram which defines the nominal residual strength of the material under fatigue is presented, and its potential applications are discussed.

2. Scheme of Experimental Study

The material tested was a commercial-grade S25C steel, which was annealed at 850°C for 1 h followed by furnace cooling. The chemical composition of the material is shown in **Table I**. **Figure 1** shows the specimen geometry and the two-step test procedures. As illustrated, each test specimen underwent a number of torsional cyclic loadings first, and then a tension test was carried out to obtain the tensile strength of the preloaded specimen. The cyclic loads were applied to propagate cracks in the specimen to investigate how the propagation of cracks affects the material strength, and how strength degradation takes place in metal under fatigue.

The reason for using torsional loads was to ensure stable crack growth in a test specimen under cyclic loading. (The previous plan was to use tension-compression cyclic loads for preloading the test specimens. After several unexpected premature failures of test specimens before reaching the designated number of load cycles, it was realised that a large number of test specimens would be required for the test. Therefore, the test plan was changed and torsional loads were used for preloading instead, which did not affect the original purpose of the research.) In order to choose a suitable stress amplitude at which to apply the cyclic loads and to determine the other testing conditions for the experiment, a torsional fatigue test was performed first to obtain the *S-N* curve, which is shown in **Figure 2**. The stress ratio of the test was -1, and the *S-N* data were obtained by employing two torsional fatigue testing machines for crosschecking. As shown in **Figure 3**, the test

Table I. Chemical composition (wt%).

C	Si	Mn	P	S	Cu	Ni	Cr	CC9
0.24	0.26	0.46	0.019	0.005	0.02	0.01	0.03	0.04

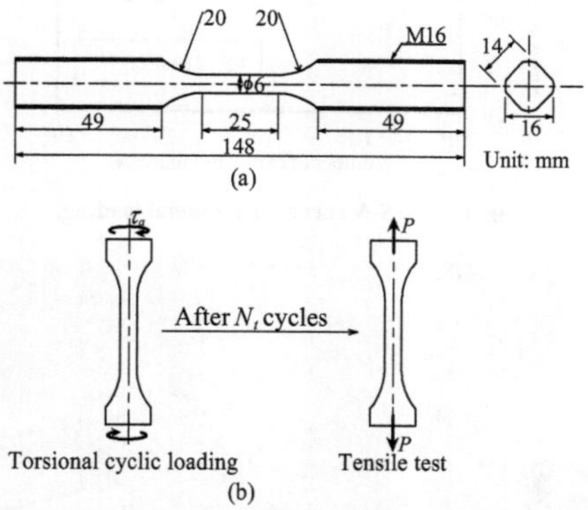

Figure 1. Specimen geometry and test procedures: (a) specimen geometry; (b) two-step test procedures.

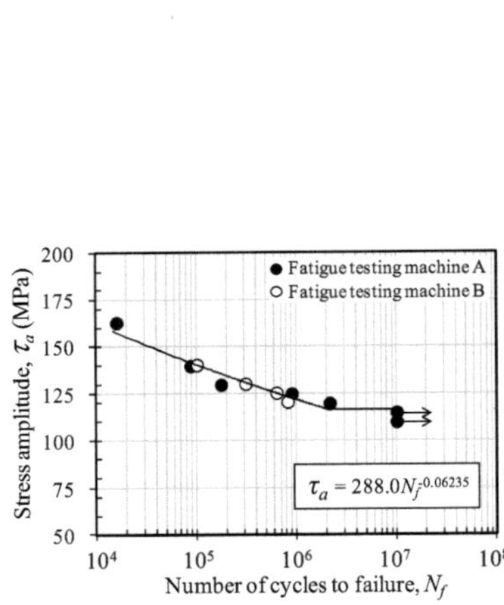

Figure 2. *S-N* curve for torsional loading.

Figure 3. Torsional fatigue test.

was carried out in open air at a rotational speed of 2000 cpm. Note that fatigue failure (N_f) was defined as the number of cycles at which the increment in the rotational amplitude reached 3.375 mm during the test, which was approximately equivalent to an increment of 1 deg. The shear stress amplitude, τ_a, is obtained as

$$\tau_a = M_a \cdot d/(2I_p) \tag{1}$$

where M_a = rotational moment amplitude; $I_p = \pi d^4/32$, polar moment of inertia; and d = diameter.

As shown in **Figure 2**, the maximum amplitude of shear stress tested was 163 MPa and fatigue failure occurred at 1.6×10^4. The fatigue limit was reached at the stress amplitude of 115 MPa, i.e., no failure at 1.0×10^7. In order to accelerate crack propagation in a test specimen to reduce the total testing time of the experiment, the stress amplitude for the detailed incremental load-cycle tests was chosen as 140 MPa ($\risingdotseq \tau_y = 143$ MPa based on the von Mises yield criterion), and the N_f of the two *S-N* data at this stress amplitude were 8.8×10^4 and 1.0×10^5, respectively. Therefore, a series of cyclic loading tests under the condition of $\tau_a = 140$ MPa was carried out from N_t of 5.0×10^4 (no cracks were observed up to this point) to N_t of 1.0×10^5 (with one exception), with a load-cycle increment of 5.0×10^3. After imposing various loading histories on the test specimens, all the specimens including those of the *S-N* tests underwent tension tests to obtain the tensile strength for each test specimen.

3. Results and Discussion

3.1 Multistage Strength Degradation

Table II shows the test piece number, test conditions and experimental results, including the crack grade, the 0.2% offset yield strength, tensile strength, elongation and contraction ratios. As seen, the severity of cracking of a test specimen after experiencing the designated number of cyclic loadings was divided into five grades: A, B, C, D and E. Typical sample cracks for each grade are shown in **Figure 4**, where Grade A = almost invisible crack (COW ≒ 0.01 mm); Grade B = visible hair-like crack (COW ≒ 0.05 mm); Grade C = clearly visible crack (COW ≒ 0.07 mm); Grade D = severe crack (COW ≒ 0.10 mm); Grade E = penetrating crack (COW ≒ 0.20 mm); and COW = Crack Opening Width. These cracks appear to be mixed-mode cracks, probably involving all three modes of deformation. Though in principle three tests were performed under each test condition, in some cases one or two extra tests were also carried out for verification. **Figure 5** shows the relationships of the tensile strength versus the load cycles imposed, using the crack grading as a principal index for the various degrees of material damage sustained by the test specimens. For clarity, the obtained strength degradation relations under τ_a = 140 MPa are shown first in **Figure 5(a)** on a linear scale, and the complete results including those of the *S-N* tests are then presented in **Figure 5(b)** on a log scale for a general view of the multistage strength degradations among all the test specimens. The average strength of each grade is represented by a solid line, and for comparison the average strength of the six virgin specimens (= 442 MPa) is shown by a dotted line.

As seen from **Figure 5**, while the average strength of grade A specimens slightly increased from that of the virgin specimens by 3%, apparently due to cyclic hardening, the average strengths of the other groups in the order of grade B to grade E decreased by 7%, 17%, 34% and 60%, respectively. The accelerated strength degradation in lower grades is conspicuous: From grade C to grade D the increase of reduction rate approaches 20%, and from D to E, it exceeds 25%. **Figure 6** shows the multistage strength degradations in the order of crack grade, which is the natural order of crack initiation and growth in a test specimen under fatigue loading. The small scatter of the individual data from the average strength in each group shows the appropriateness of using the present crack grading method for clarifying multistage strength degradation. This indicates that, in essence, each crack grade represents a unique stage of crack propagation in the material, and therefore a change of crack grade implies a change of cracking behaviour.

With the constant stress amplitude of τ_a = 140 MPa, the strength of the specimens endured a complete transition of crack grade, i.e., from A to E, even though the grade E test data were mainly composed of test specimens with smaller stress amplitudes of 120 to 130 MPa and in a much higher cyclic loading range. As expected, the test specimen with the highest stress amplitude of 163 MPa and the lowest loading cycles of 1.6×10^4 also fell into grade E. In fact, almost all the strength data of grade E were obtained from the specimens of the *S-N* tests, and in this sense grade E is called the failure grade.

It is worth noting that among the grade A specimens, the tensile strengths of the two *S-N* test specimens at the fatigue limit remained virtually the same as those of the virgin specimens; the specimens showed no evidence of cyclic hardening due to the much smaller stress amplitudes endured during cyclic loading. The experimental results in **Figures 5** and **6** confirm the applicability of the multistage strength degradation theory to metal fatigue, which in principle cannot be verified directly on the same test specimen by experiments (obviously, because the ultimate strength can only be obtained by actually fracturing the test specimen itself). By testing multiple identical specimens under varying cyclic loading conditions, however, the general trend and mechanism of strength degradation can be established.

Table II. Test conditions and experimental results of ultimate tensile strength.

	τ_a (MPa)	N_f(*), N_t (cycle)	Crack grade	0.2%σ_y (MPa)	σ_u (MPa)	Elng (%)	d/D (%)
1	-	-	-	246	446	31.5	55.3
2	-	-	-	244	444	33.5	54.8
3	-	-	-	259	440	36.0	55.9
4	163	1.6×10^4*	E	110	151	5.5	-
5	120	2.1×10^6*	D	237	313	12.5	5.4
6	140	8.8×10^4*	D	199	271	8.5	11.8
7	110	1.0×10^7	A	220	441	35.5	52.8
8	130	1.8×10^5*	B	252	388	14.5	17.6
9	125	9.0×10^5*	E	147	171	3.5	-
10	115	1.0×10^7	A	225	447	32.0	52.6
11	130	3.2×10^5*	E	-	178	6.0	-
12	125	6.4×10^5*	E	125	210	8.0	1.6
13	140	1.0×10^5*	B	246	407	13.0	14.3
14	120	8.3×10^5*	E	138	192	3.5	2.4
15	140	6.1×10^4	B	266	407	16.0	20.8
16	140	6.5×10^4	B	251	434	18.0	20.6
17	140	5.0×10^4	A	263	463	32.0	47.6
18	140	5.0×10^4	A	259	459	30.0	47.8
19	140	5.0×10^4	A	255	458	38.0	52.7
20	140	7.3×10^4	B	258	419	19.0	29.6
21	140	6.0×10^4	A	257	462	32.0	54.3
22	140	6.0×10^4	A	253	454	33.0	54.6
23	140	1.3×10^5	D	206	287	15.0	7.3
24	140	7.6×10^4	A	271	444	17.0	26.3
25	140	7.0×10^4	A	257	456	32.5	50.5
26	140	7.0×10^4	A	254	458	32.5	49.8
27	140	7.0×10^4	A	257	457	29.0	48.0
28	140	8.2×10^4	B	260	394	11.5	13.3

Note: **Elng, percent elongation; *d/D*, contraction percentage**.

(Continued)

Table II. Test conditions and experimental results of ultimate tensile strength (continued).

TPNo	τ_a (MPa)	N_f(*), N_t (cycle)	Crack grade	0.2%σ_y (MPa)	σ_u (MPa)	Elng (%)	d/D (%)
29	140	8.0×10^4	A	249	458	34.5	52.9
30	140	6.5×10^4	B	279	394	12.0	17.4
31	140	8.0×10^4	A	245	458	34.0	52.4
32	140	8.0×10^4	A	269	458	21.0	34.1
33	140	8.0×10^4	A	254	464	30.0	52.4
34	140	6.2×10^4	C	248	369	8.0	11.6
35	140	7.8×10^4	C	257	373	12.0	13.7
36	140	7.5×10^4	A	258	455	27.5	37.5
37	140	7.5×10^4	A	252	461	31.0	52.3
38	140	7.5×10^4	A	257	459	28.5	54.5
39	140	7.5×10^4	A	253	456	33.5	54.3
40	140	6.5×10^4	A	252	458	33.0	52.1
41	140	6.5×10^4	A	252	458	27.0	52.4
42	140	6.5×10^4	A	250	456	34.5	52.3
43	140	6.5×10^4	A	255	457	32.0	51.7
44	140	6.0×10^4	A	264	452	19.5	29.0
45	140	6.0×10^4	A	254	462	30.0	52.4
46	140	8.5×10^4	B	261	411	15.5	14.6
47	140	8.5×10^4	E	161	166	4.5	1.4
48	140	8.5×10^4	C	253	373	12.0	14.1
49	140	8.5×10^4	C	230	366	12.5	11.7
50	140	9.0×10^4	C	236	363	10.0	8.4
51	140	9.0×10^4	B	262	436	19.5	25.7
52	140	8.0×10^4	B	270	425	15.0	17.7
53	140	9.0×10^4	C	254	362	4.0	4.1
59	-	-	-	250	439	36.5	55.9
60	-	-	-	235	434	37.5	55.9
61	-	-	-	250	447	33.0	53.5

Note: **Elng, percent elongation; *d/D*, contraction percentage.**

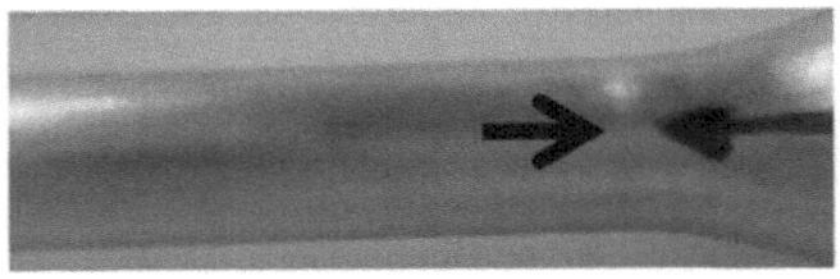

Grade A

Grade B

Grade C

Grade A: Almost invisible crack (COW ≒ 0.01 mm)
Grade B: Visible hair-like crack (COW ≒ 0.05 mm)
Grade C: Clearly visible crack (COW ≒ 0.07 mm)
Grade D: Severe crack (COW ≒ 0.10 mm)
Grade E: Penetrating crack (COW ≒ 0.20 mm)

(Continued)

Figure 4. Five grades for classifying the degree of severity of cracking in test specimens with cyclic loading history

Microscopically, metal is a heterogeneous material. Under repeated loading, material damage starts from the weakest points and will continue until most of the weak linkages in the material structure, which form the first layer (or level) of resistance, have been destroyed. Then, the damage process moves to the next stage, destroying the next layer of material components in a similar fashion, and the layer after that. At each stage of material damage or crack propagation, as the overall material structure remains the same, only small variations in the ultimate strength are expected. However, advancing from one stage to the next will fundamentally change and weaken the overall material structure and cause a substantial reduction of the ultimate strength. The broad destruction of the material structure layer by layer under repeated loads contrasts sharply with the localized damage sustained by each of the same layers under a monotonically-increasing load. These in-depth analyses of the unique material failure processes under cyclic loading help explain why the tensile strengths of grade B to grade E in **Figures 5** and **6** are approximately constant, and why strength degradation takes place intermittently.

3.2 Strain Energy Reduction

Figure 7 shows the load-displacement relations of all the test specimens obtained by tensile tests. Based on these relations, the total strain energy at the maximum load and at the failure load

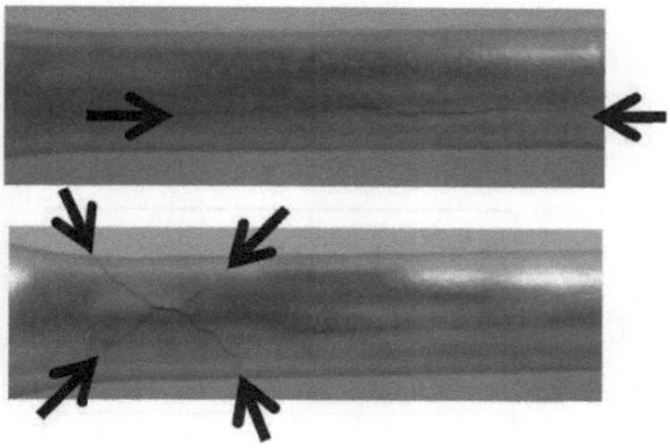

Grade D

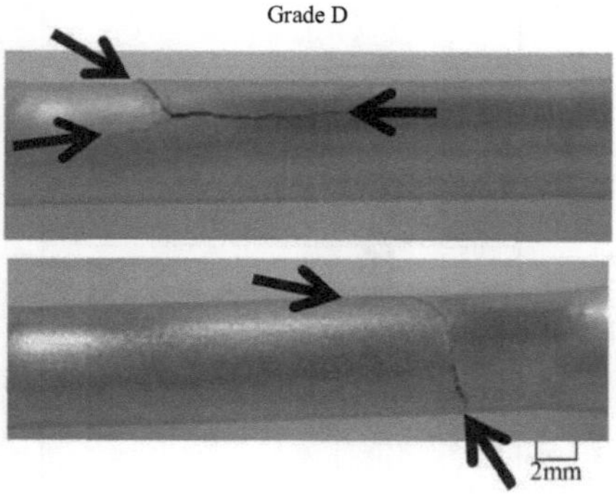

Grade E

Grade A: Almost invisible crack (COW ≒ 0.01 mm)
Grade B: Visible hair-like crack (COW ≒ 0.05 mm)
Grade C: Clearly visible crack (COW ≒ 0.07 mm)
Grade D: Severe crack (COW ≒ 0.10 mm)
Grade E: Penetrating crack (COW ≒ 0.20 mm)

Figure 4. **Five grades for classifying the degree of severity of cracking in test specimens with cyclic loading history (continued).**

were calculated separately for each preloaded specimen and the results are shown in **Figure 8(a)** and **8(b)**, respectively. Following the example of **Figure 6**, the same results are also shown in the order of crack grading in **Figure 9(a)** and **9(b)**. As the severity of cracking progresses and the material becomes fragile, the total strain energy decreases rapidly and the two energy states become similar.

Figure 10 shows the ratio of the average total strain energy at the maximum load to that of the virgin specimens for each crack grade. Compared with the relatively large value of 0.87 for grade A, the rapidly decreasing ratios of the other grades, which were 0.38, 0.23, 0.17 and 0.05 from B to E, are noteworthy. Indeed, these ratios contrast sharply to the relatively slow rates of strength reduction discussed previously, which were 7%, 17%, 34% and 60% down from the average strength of the virgin specimens in the order of B to E. As these ratios reflect the decrease of the material's deformation capability, they indicate how brittle the test specimens became at each stage of crack propagation. Hence, when assessing material properties under fatigue, the strength reduction and the loss of strain energy should both be taken into consideration. Among the five grades, the 60% strength

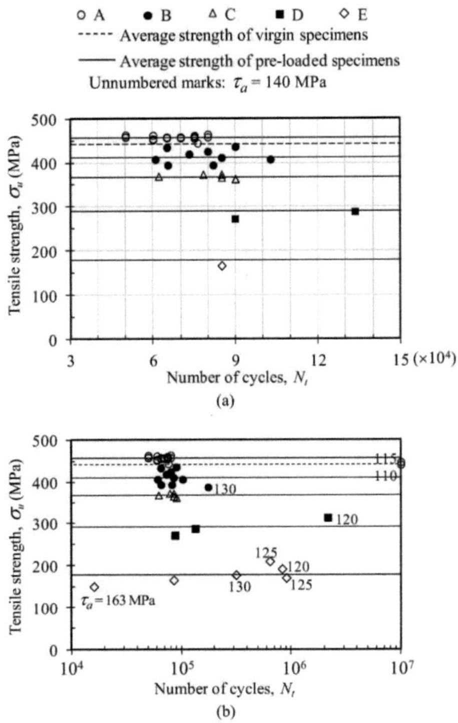

Figure 5. **Multistage strength degradation based on the advance of crack grade: (a) τ_a = 140 MPa; (b) all test specimens [16].**

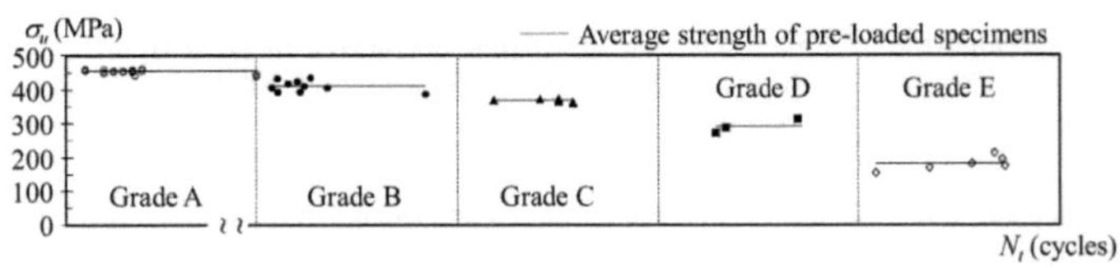

Figure 6. **Multistage strength degradation based on the order of crack grade (load cycles in the local scale of each grade).**

reduction and the 95% loss of deformation energy for grade E, the failure grade, should be emphasised because these values reveal the fragile material structures of the test specimens prior to failure and clearly explain why fatigue failure is inevitably brittle.

3.3 A Partially Modified S-N Curve Based on Grade D

It is well-known that *S-N* tests show a wide dispersion of data that increases exponentially as the stress amplitude decreases and approaches the fatigue limit. The scatter of the test results can be

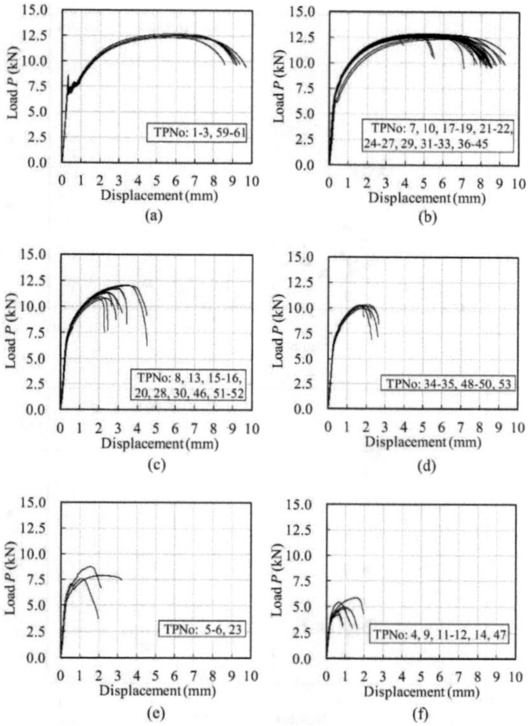

Figure 7. Load-displacement relations of test specimens: (a) virgin specimens; (b) Grade A; (c) Grade B; (d) Grade C; (e) Grade D; (f) Grade E.

attributed to various random factors, most of which are related to the micro differences in material structure and geometrical surface of the test specimens [17, 18]. This unique feature of fatigue tests can be understood by considering the weak and brittle material properties of the specimens in lower grades, and among them grade E should be singled out as an extreme. Unlike the material properties of grades C and D, a grade E specimen has no redundant strength and stiffness left in its severely-damaged material structure and thus, any random factor that either accelerates or slows crack propagation in the material can immediately influence the outcome of the test, contributing to the large scatter in the obtained fatigue life, N_f.

Figure 11(a) shows an *S-N* curve with five sections of descending strength grades, beginning with grade A for the first section and ending with grade E for the last section including the fatigue limit. Obviously, the *S-N* curve in any one of the sections represents the occurrence of material failure, N_f, when the transition of strength grade to the next lower level takes place under the cyclic stress, *S*. Now, a fundamentally different approach to the *S-N* curve in the last region has emerged from the present study, i.e., to redefine the curve as the number of load cycles required for the strength degradation of the material to reach grade D, as shown in **Figure 11(b)**. As a result, the fatigue limit is lowered. As seen, the *S-N* curve above the region of grade E remains the same, still representing material failure. However, the modified *S-N* curve in the last region no longer represents fatigue

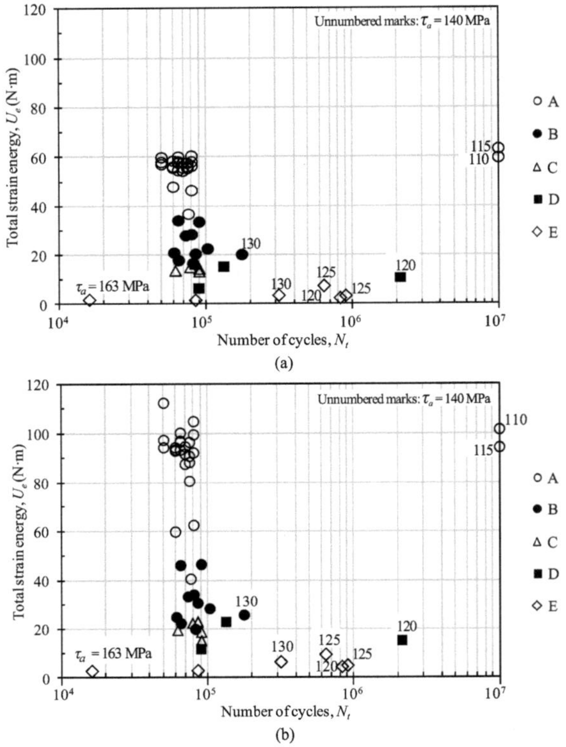

Figure 8. **Total strain energies of test specimens: (a) at the maximum load; (b) at the failure load.**

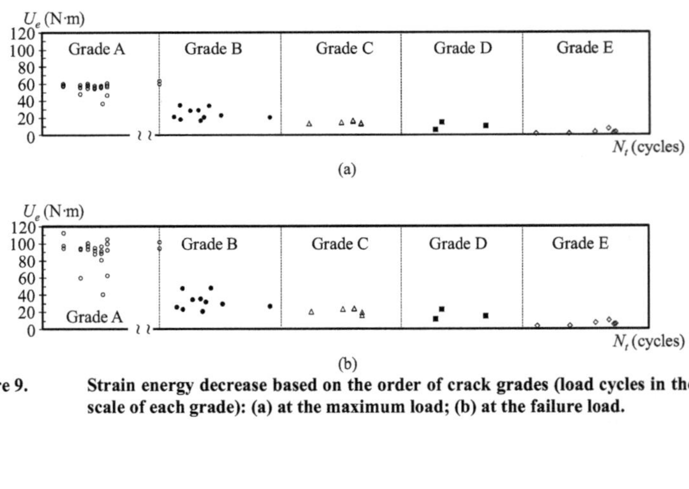

Figure 9. **Strain energy decrease based on the order of crack grades (load cycles in the local scale of each grade): (a) at the maximum load; (b) at the failure load.**

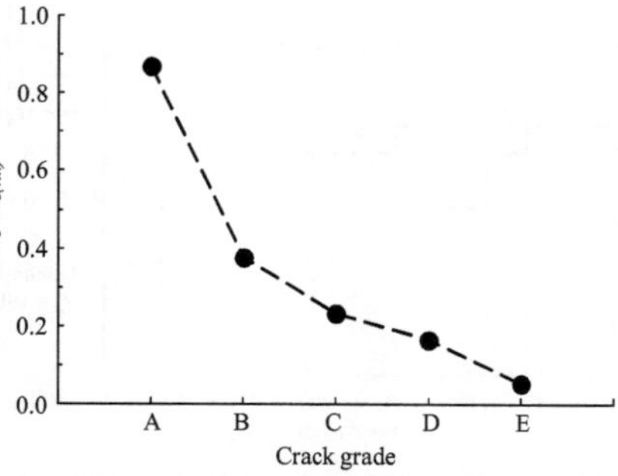

Figure 10. **Ratio of the averaged total strain energy at the maximum load to that of the virgin specimens with respect to crack grade.**

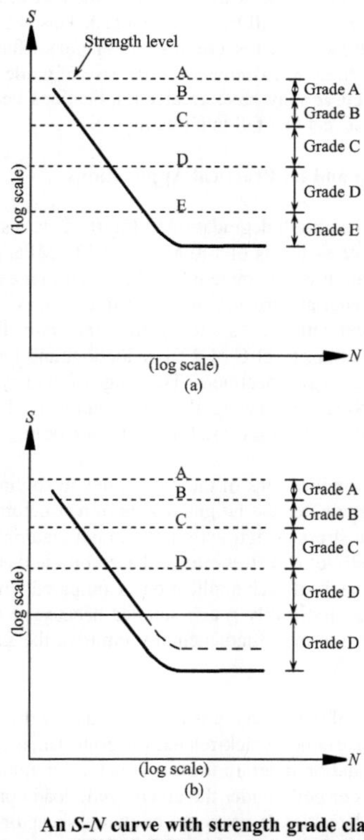

Figure 11. **An *S-N* curve with strength grade classifications and a modified *S-N* curve: (a) an *S-N* curve with five strength grades; (b) a modified *S-N* curve removing Grade E and redefining the last section based on Grade D.**

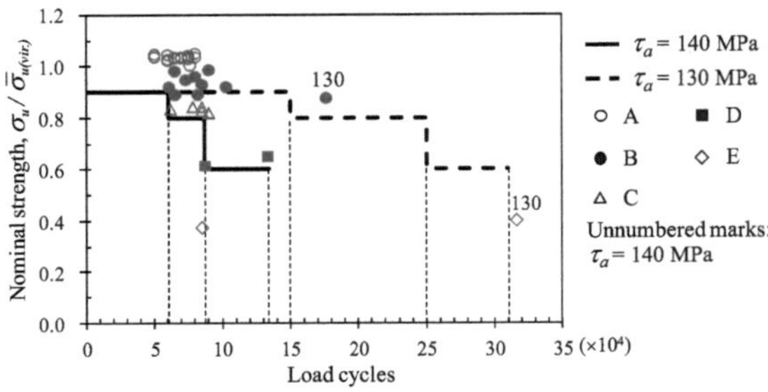

Figure 12. **Strength degradation diagrams at two stress levels.**

failure. Instead, it indicates the material degradation to grade D at different stress levels of the cyclic loads. Although random factors in fatigue tests still exist, their effects on the test result, i.e., the number of load cycles to reach grade D, will be much reduced. This is because the partially-damaged material structure of grade D still has some reserve of strength and stiffness, and therefore its response to random factors will not be as direct or extreme as in the case of grade E. It is worth noting that this partial modification to the *S-N* curve may have greater implications because fatigue design is often performed based on small stress levels that fall into this region.

Strength Degradation Diagram and Its Practical Applications

Based on the test results of strength degradation in **Figure 5**, two strength degradation diagrams are drawn in **Figure 12** for the stress levels of 140 MPa and 130 MPa, respectively. These diagrams define the nominal residual strength of the material in the given range of load cycles, providing the required but often-unknown material strength in the safety check and evaluations of existing structures or machine parts under fatigue. As shown, there are only three nominal strength levels considered, level B of 10% reduction, level C of 20% reduction, and level D of 40% reduction from the average tensile strength of the virgin specimens. The range of load cycles for each strength level is determined from test results. Note that due to the small number of tests carried out, especially regarding the case of 130 MPa, these diagrams are for illustration only.

There are several possible applications of these diagrams in solving fatigue-related engineering problems. The first apparent usage is for the fatigue design of a structural member or a machine part, which in essence is an allowable stress design with the nominal material strength being derived from one of these diagrams with a well-defined stress range, based on the design life. The allowable stress design method is the theoretical basis of such applications. Compared with the conventional *S-N* curve, the scatter in such a relation is predictably much smaller because it is obtained under stable test conditions, and hence its application may fundamentally improve the safety and economy of fatigue design.

The second possible usage of these diagrams is for evaluating the safety of a structural member (or a machine part), which has sustained crack-related material damage under fatigue loads. In such cases, a complete strength degradation diagram for this structural member should be established first, by tests or by numerical analysis or both, under the given cyclic load condition. The member strength in the diagram may represent the ultimate sectional moment, shear or axial force of that member, depending on the structural mechanics nature of the problem. These strength degradation relations are predictably multistage, because the member strength is derived from the material strength, which decreases through multiple stages under cyclic loading. Obviously, crack grades for the given structural member should be properly classified, with each grade uniquely related to a specific strength level in the multistage member-strength degradation.

It should be pointed out that, in general, the typical cracking behaviour observed in a small test specimen under simple load conditions for basic material property tests is fundamentally different from that found in a much larger structural member or a machine part under complex load conditions, in terms of the crack scale, number of cracks, etc. Hence, the crack grades for the present structural member should not be confused with the crack grades defined in **Figure 4** for small test specimens under torsional cyclic loads.

By numerical analysis, the critical member force in the cracked structural member can be determined, which is then checked against the nominal member strength derived from the strength degradation diagram, based on the fatigue life and the crack grade rated for the actual cracks that occurred in that structural member. Finally, a decision on whether this damage-sustained structural member is safe or not can be reached by applying the limit state design method, which provides a solid theoretical basis for evaluating the safety of various aging structures or machine parts with fatigue cracks.

4. Conclusions

An experimental investigation on the fatigue mechanism of metal was reported. In this test study, multiple identical test specimens were first preloaded with torsional cyclic loads of increasing number of load cycles, and then tension tests were carried out to obtain their tensile strengths. The following conclusions were drawn:

1. The experimental results revealed the existence of five distinct stages of crack propagation and strength degradation in S25C steel under the given test conditions, and the rates of strength reduction from the second to the fifth stage were 7%, 17%, 34% and 60%, respectively. The study confirmed the applicability of the multistage strength degradation theory as the fatigue mechanism of metal.
2. Examinations of surface cracks showed that each stage of crack propagation was associated with a unique cracking behaviour on the surface of a specimen, indicating the sudden change of cracking behaviour at some critical point of crack growth in metal as the cause for strength degradation to take place.
3. Compared with strength degradation, the reduction of total strain energy in a test specimen was more prominent. For grade E, the failure grade, the strength reduction was 60% and the loss of deformation energy reached 95%. These extremely weak and brittle material properties of the failure grade help explain the wide data dispersion in fatigue tests under small stress amplitudes.
4. To improve the theoretical basis of fatigue design by reducing data dispersion in fatigue tests, a modified *S-N* curve for the region of small stress amplitudes was proposed. The grade-D-based section of the curve represents the number of load cycles required for the strength degradation of a material to reach grade D. Due to the remaining strength and stiffness in grade D, data scatter in the new curve is expected to be small.
5. To facilitate maintenance work on existing structures and machine parts, a strength degradation diagram which defines the nominal residual strength of a material in the given range of load cycles was proposed. Potential applications of the diagram for the fatigue design of a structural member and safety evaluations of existing systems with fatigue cracks were discussed.

Acknowledgment

This research was carried out in collaboration with the Amagasaki unit of Nippon Steel & Sumikin Technology Co., Ltd., who performed the tests reported in this study.

References

1. W. A. J. Albert, Archive für Mineralogie, Geognosie, Bergbau und Hüttenkunde, 10, 215-34 (1838).
2. A. Wöhler, Zeitschrift für Bauwesen, 10 (1860); English summary: Engineering, 4, 160-61 (1867).
3. W. Fairbairn, **Phil. Trans. Roy. Soc.**, 154, 311 (1864).
4. H. Gerber, Zeitschrift des Bayerischen Architeckten und Ingenieur-Vereins, 6, 101-10 (1874).
5. J. Goodman, Mechanics Applied to Engineering, Longmans Green, London (1899).
6. J. A. Ewing, and J. C. Humfrey, **Phil. Trans. Roy. Soc.,** A200, 241-50 (1903).
7. O. H. Basquin, Proceedings of the American Society for Testing and Materials, 10, 625-30 (1910).
8. A. A. Griffith, **Phil. Trans. Roy. Soc.**, A221, 163-97 (1921).
9. W. Weibull, Proceedings 151, Royal Swedish Academy of Engineering Sciences (1939).
10. G. R. Irwin, **Journal of Applied Mechanics**, 24, 361-64 (1957).
11. D. S. Dugdale, **Journal of the Mechanics and Physics of Solids**, 8, 100-08 (1960).
12. P. C. Paris, M. P. Gomez, and W. P. Anderson, The Trend in Engineering, 13, 9-14 (1961).
13. J. R. Rice, Fatigue Crack Propagation: Special Technical Publication, 415 (Philadelphia; American Society for Testing and Materials), 247-309 (1967).
14. Z. Shi, Crack Analysis in Structural Concrete: Theory and Applications, Butterworth-Heinemann, Burlington, MA, p 149-78 (2009).
15. Z. Shi, Y. Nakamura, and M. Nakano, Int. J. Fatigue, 33, 1140-50 (2011).
16. Z. Shi, M. Nakano, and C. Liu, 11th International Fatigue Congress; Advanced Materials Research, 891-892, 753-58 (2014).
17. E. Zahavi, and V. Torbilo, Fatigue Design: Life Expectancy of Machine Parts, CRC Press, p 58 (1996).
18. J. Schijve, Fatigue of Structures and Materials, Kluwer Academic Publishers, p 310-28 (2001).

Fatigue of Materials III: Advances and Emergences in Understanding
Edited by: T.S. Srivatsan, M. Ashraf Imam, and R. Srinivasan
TMS (The Minerals, Metals & Materials Society), 2014

THE STRESS CONTROLLED CYCLIC FATIGUE AND FRACTURE BEHAVIOR OF ALLOY STEEL 300M

K. Manigandan[1], T.S. Srivatsan[1], G. Doll [2], and T. Quick[3]

1. *Department of Mechanical Engineering*
3. *Department of Geology*
The University of Akron
Akron, Ohio 44325-3903, **USA**
E-Mail: tss1@uakron.edu

2. ***Timken Professor*** *of Surface Engineering*
Director, Timken Engineered Surfaces Laboratories
Akron Engineering Research Center
The University of Akron, USA
Akron, OH 44325-0406

Abstract

In this manuscript the results of a study aimed at understanding the extrinsic influence of test specimen orientation, with respect to wrought alloy steel plate, on high cycle fatigue properties and fracture behavior is highlighted. The alloy steel chosen was 300 M. Samples of this alloy steel prepared from both the longitudinal and transverse orientation were cyclically deformed over a range of maximum stress and the corresponding number of cycles to failure (N_F) was recorded. The influence of test specimen orientation and intrinsic microstructural effects on cyclic fatigue life is presented. At the chosen test temperature, the macroscopic fracture mode was essentially identical regardless of the orientation of the test specimen with respect to the wrought plate. The microscopic mechanisms governing cyclic deformation, fatigue life and final fracture behavior is presented in light of the mutually interactive influences of magnitude of applied stress, intrinsic microstructural effects, orientation of test specimen, and deformation characteristics of the constituents in the microstructure of this alloy steel.

Key words: Alloy steel, orientation, microstructure, cyclic fatigue

1. Introduction

The continuing need arising from the sustained demand for new and improved materials for use in a wide variety of applications spanning the domains of performance-critical, non-performance critical, and even house-hold items to include sporting goods has certainly provided the much needed initiative, incentive and impetus for engineering sustained developments that eventually culminated in the emergence of a wide variety of materials to include the family of metal alloys for a specific application. An attractive combination of high strength, high elastic modulus, high specific strength (σ / ρ), superior corrosion resistance coupled with novel and noticeable innovations in processing techniques spanning the domains of both primary and secondary has made possible the ease of manufacturing specialty steel, which does offer acceptable combinations of high strength, moderate ductility, adequate fracture toughness and acceptable fatigue resistance [1-6]. In recent years, alloy steel 300M has been chosen for use in those applications that require high strength, in the range 290-300 ksi, such as, (i) aircraft landing gear, (ii) high strength bolts, and even (iii) parts of airframe.

Often, the fatigue strength of a low carbon steel or medium carbon steel increases with an increase in ultimate tensile strength [σ_{UTS}]. However, for the case of specialty steel whose ultimate tensile strength [σ_{UTS}] exceeds 1800 MPa, the fatigue strength of the material would depend on both its strength and toughness. The microstructure of vacuum, arc-re-melted (VAR) steels is likely to be refined when compared one-on-one with the conventionally processed counterpart. The improvement in microstructure does result in noticeable improvements in strength, ductility, toughness, cyclic fatigue resistance and concomitant fracture behavior [7-12]. The noticeably high yield strength of specialty steels provides the potential for an observable improvement in mechanical performance coupled with a reduction in weight; two key factors that are of both importance and need in selecting a specialty steel for use in a spectrum of applications spanning a wide range of industries. Often, in the selection and eventual use of a low alloy high strength steel for a specific application, the tendency for failure due to fatigue is exacerbated by the initiation of fatigue cracks at the large metallic inclusions present in the microstructure [13-16]. However, crack initiation at an inclusion can be suppressed when size of the inclusion is less than 5-microns, such as in the case of steels that have a "clean" microstructure. Similarly under conditions of cyclic fatigue there occurs a critical size of the inclusion below which initiation of failure due to cracking of the inclusion will not be easily favored to occur. The specific role of microstructural constituents and intrinsic microstructural effects in governing the fatigue behavior spanning the domains of:

(a) Frequency effects on ultrasonic fatigue response of ultra-fine-grained steels [17],
(b) Combined use of experimental testing and J-integral analysis for studying crack growth behavior in an alloy steel [18], and
(c) Fatigue properties of a bainite/martensite dual phase high strength steels [19],

are few of the recent studies on alloy steels that is documented in the "open" literature.

The objective of this paper is to present and discuss the results of a study aimed at investigating the influence of load ratio and test specimen orientation on stress-controlled high cycle fatigue response and resultant fracture behavior of a high strength alloy steel. The chosen steel was cyclically deformed over a range of maximum stress at a given load ratio. The conjoint influence of load ratio and microstructure on fatigue life (N_f) and resultant fracture behavior is discussed following scanning electron microscopy observations of the cyclically deformed and failed specimens. The primary purpose of this research exercise was to concurrently evaluate the specific role of microstructural constituents and intrinsic microstructural effects in governing fatigue life (N_f) and fracture behavior of the chosen high strength alloy steel at a specific load ratio.

2. Material

The high strength alloy steel chosen for this study was 300M (UNS K44220), provided by Latrobe Specialty Steel Company (Latrobe, PA, USA). The chemical composition (in weight percent) of this high strength alloy steel is provided in **Table 1.** The presence of carbon in substantial amount in this alloy steel facilitates solid solution strengthening besides increasing hardenability through the formation, presence and distribution of alloy carbides. Presence of the elements chromium (Cr) and molybdenum (Mo) results in the formation and presence of the second-phase particles chromium carbide (Cr_2C_3) and molybdenum carbide (Mo_2C), which are dispersed through the microstructure. The carbide particles distributed through the microstructure enhance the strength of the steel matrix [20, 21]. However, the presence and distribution of a substantial number of carbide particles in the microstructure is detrimental to ductility, fracture toughness and fatigue crack initiation resistance. This can be attributed to a 'local' increase in the number of sites that are easily susceptible to the early initiation of fine microscopic cracks. Presence of the element nickel (Ni) in the alloy steel helps in lowering the transition temperature thereby making the steel receptive for use at low service temperatures. The chosen steel, i.e., 300 M, contains an appreciable amount of molybdenum (0.40 pct.), which aids in refining the grain size besides forming molybdenum carbide (Mo_2C) and its resultant influence in enhancing toughness. This steel was produced using the method of Arc VAR (VAR: Vacuum Arc Remelting; with Arc being an initial arc-melt operation) to provide optimum cleanliness and preferred ingot structure.

Table 1 **Nominal chemical composition of the Alloy Steel 300M .**

C	Mn	Si	Ni	Cr	Mo	V
0.42	0.75	1.65	1.80	0.80	0.40	0.07

3. Experimental Procedures

3.1 Initial Characterization of Microstructure

A sample of the chosen high strength steel was prepared very much in conformance with standard procedures used for metallographic preparation of metal samples. This involved a coarse polish of the as-cut sample of the candidate steel using progressively finer grades of silicon carbide (SiC) impregnated emery paper (i.e., 320-grit, 400-grit, and 600-grit) followed by fine polishing using 5 micron and 1.0 micron alumina-based polishing compound suspended in distilled water as the lubricant. The as-polished sample was subsequently etched using nital, i.e., a solution mixture of nitric acid in methanol. Etching helped reveal the following: (i) grain boundaries, (ii) morphology of the grains, and (iii) other intrinsic features, such as, (a) location, (b) morphology, and (c) distribution of both the coarse and intermediate-size second-phase particles in the microstructure. The polished and etched sample was examined in an optical microscope, at low magnifications, and photographed using standard bright field illumination technique.

3.2 Test Specimen Preparation

Tensile and fatigue test specimens were precision machined from the as-provided plate stock of this alloy steel. The test specimens, were machined such that the major stress axis was parallel to both the longitudinal and transverse directions of the as-provided plate stock. The cylindrical test specimens conformed well to the specifications outlined in the standard ASTM E-8 and had threaded ends. At the gage section, the test specimen measured 3.175 mm in diameter and 12.5 mm in length. To minimize the effects and/or contributions from surface irregularities and finish, final surface preparation was achieved by mechanically polishing the gage section of the machined test specimens to remove all of the circumferential scratches and surface machine marks.

3.3 Mechanical Testing

The mechanical tests, to include both tensile and cyclic fatigue, were performed on a fully-automated, closed-loop servo-hydraulic mechanical test machine [Model: INSTRON-8500 Plus] using a 100 KN load cell. All of the tests were conducted in the room temperature (300 K) laboratory air (Relative Humidity of 55 pct.) environment. For the uniaxial tensile test the test specimens were deformed at a constant strain rate of 0.0001/sec. An axial 12.5-mm gage length clip-on type extensometer was attached to the test specimen at the gage section using rubber bands. The stress and strain measurements, parallel to the load line, and the resultant mechanical properties were provided as an output by the control unit of the test machine.

The value of yield strength obtained from the uniaxial tensile test was used to calculate the following:

(a) the yield load, and

(b) maximum stresses as a percentage of the yield stress of the alloy, for the purpose of testing specimens of this high strength alloy steel in fatigue under conditions of stress amplitude-control.

The total stress amplitude-controlled high-cycle fatigue (HCF) tests were performed using a sinusoidal waveform and at stress ratio [R = minimum stress / maximum stress] of 0.1. The appropriate load ratio was obtained in cyclic tension with a positive mean stress (R = 0.1). The stress amplitude- controlled fatigue tests were conducted at a constant cyclic frequency of 5 Hz. At each chosen stress ratio [$R = \sigma_{minimum} / \sigma_{maximum}$], the fatigue tests were conducted over a range of stress amplitudes ($\sigma_{maximum}$ - $\sigma_{minimum}$ /2) to establish the variation of fatigue life (N_f) with maximum stress ($\sigma_{maximum}$). The test data collected from the stress-amplitude-controlled test are used to establish the following relationships:

(i) Variation of fatigue life (N_f) with maximum stress [$\sigma_{maximum}$].
(ii) Variation of fatigue life (N_f) with maximum elastic stain [$\varepsilon_{maximum} = \sigma_{maximum} / E$]
(iii) Variation of fatigue life (N_f) with the ratio of ($\sigma_{maximum} / \sigma_{yield}$)
(iv) Variation of fatigue life (N_f) with the ratio of [$\sigma_{maximum} / \sigma_{UTS}$].

The relationship was determined for test specimens that were cyclically deformed at a stress ratio of 0.1. The specific influence of test specimen orientation with respect to the as-provided plate stock can be inferred from the trend shown by the relationship.

Table 2 **A compilation of the room temperature tensile properties of 300M (T = 25°C). [Results are the mean values of duplicate tests]**

300M	Elastic Modulus		Yield Strength		UTS		Elongation GL=0.5" (%)
	Ksi	GPa	ksi	MPa	Ksi	MPa	
Longitudinal	30751.05	212.0	64.86	447.19	130.06	897.02	15.0
Transverse	31505.09	217.0	78.93	544.08	133.35	922.02	17.3

3.4 Failure-Damage Analysis

Fracture surfaces of the cyclically deformed and failed specimens were comprehensively examined in a scanning electron microscope (SEM), over a range of allowable magnifications, to help establish the following:
(A) the macroscopic fracture mode, and
(B) characterize the fine scale topography and intrinsic features on the fatigue fracture surface with the purpose of establishing the microscopic mechanisms contributing to failure by fracture.

Samples for observation in the SEM were obtained from the cyclically deformed and failed test specimens by sectioning parallel to the fracture surface.

4. Results and Discussion

4.1 Initial Microstructure.

The optical microstructure of this high strength alloy steel, denoted as Lescalloy R 300M-HS, is shown in **Figure 1**. The observed microstructure is typical of high strength steels in that it essentially reveals a combination of carbon-rich region and a carbon-depleted region. A higher carbon and alloy content in this high strength alloy steel resulted in a greater volume fraction of martensite in the carbon-rich regions. The presence and overall morphology of the martensite was much finer and intermingled with pockets of ferrite-rich regions, i.e., the carbon depleted region in this high strength alloy steel that was produced by vacuum arc re-melting (VAR). The presence of the hard and essentially elastic deforming martensite micro-constituent is a preferred site for high 'local' stress concentration during deformation. In addition to these two phases, or micro-constituents, i.e., ferrite and martensite, dispersed through the microstructure were the carbide particles. These particles result as a consequence of the high alloy content in this high strength steel. The morphology of the martensite micro-constituent in the carbon rich region is governed by both chemical composition and the processing technique, to include both primary processing and secondary processing, used to engineer the starting material. i.e., high strength alloy steel. These two phases along with the carbide particles exert an appreciable influence on (i) cyclic fatigue response, (ii) fatigue life, and (iii) final fracture behavior at the fine microscopic level, when test specimens of this alloy steel are subject to cyclic loading.

4.2 Tensile Properties

The ambient temperature tensile properties of this high strength alloy steel for the two orientations, longitudinal and transverse, are summarized in **Table 2**. Results are the mean values based on duplicate tests. The elastic modulus of this high strength alloy steel was 212 MPa in the longitudinal orientation and 217 MPa in the transverse orientation. The yield strength of the chosen steel was 477 MPa in the longitudinal orientation and 544 MPa in the transverse orientation. In other words, the yield strength in transverse orientation was 15 percent greater than in the longitudinal orientation. For both the longitudinal and transverse orientation the ultimate tensile strength was higher than the yield strength indicating the occurrence of strain hardening beyond yield. The ultimate tensile strength in the longitudinal orientation (896 MPa) was five percent lower than in the transverse orientation (922 MPa). The ductility quantified by elongation over a gage length of 12.5 mm was 15 percent in the longitudinal orientation and 17 percent in the transverse orientation. Overall, the chosen high strength alloy steel revealed better combination of both strength and ductility in the transverse orientation than in the longitudinal orientation. The engineering stress versus engineering strain curves for the two orientations, i.e., longitudinal and transverse, of this alloy steel are shown in **Figure 2**. At and beyond the onset of necking at the maximum load this steel exhibited large plastic strain prior to failure by fracture for both the longitudinal and transverse orientations.

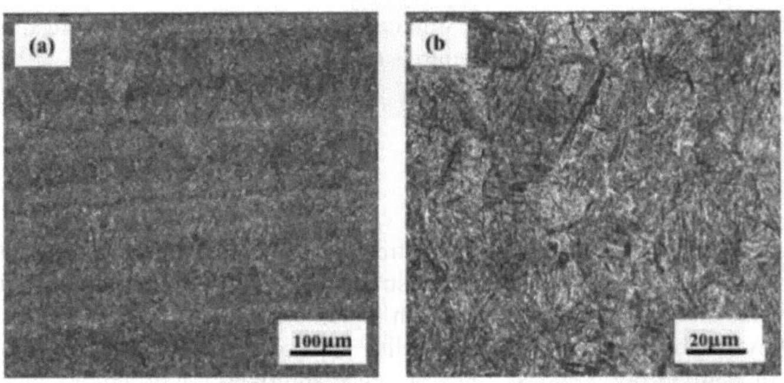

Figure 1. Optical micrographs showing the key micro-constituents of 300M at two different magnifications

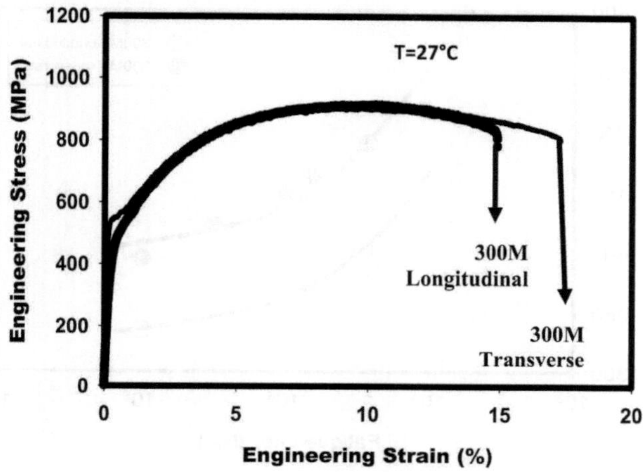

Figure 2. Influence of test specimen orientation (longitudinal versus transverse) on engineering stress versus engineering strain response of alloy 300M when deformed at room temperature (27 C).

4.3 Cyclic Stress-Amplitude Controlled Fatigue Response

At room temperature the test results of the stress amplitude-controlled fatigue tests for both the longitudinal and transverse orientations are shown in **Figure 3** as the variation of fatigue life (N_f) with maximum stress ($\sigma_{max.}$). On account of the limited amount of starting material chosen for this specific study on tensile deformation, cyclic fatigue and fracture behavior of this specific steel only one test specimen was cyclically deformed with care and caution at each chosen stress level. The maximum stress (MPa) versus fatigue life (cycles) curves show a trend of gradual increase in fatigue life (N_f) with progressive decrease in maximum stress ($\sigma_{maximum}$), a trend that is shown by most ferrous metals, their alloy counterparts and even their composite counterparts. At equivalent values of maximum stress the transverse specimen revealed 100 percent improvement in fatigue life over the longitudinal counterpart.

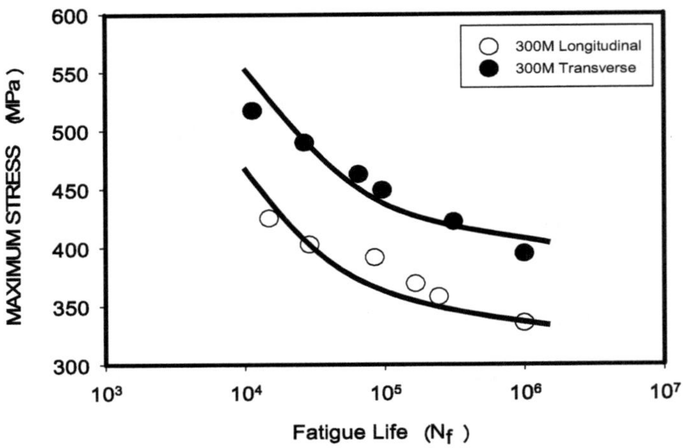

Figure 3. **Influence of test specimen orientation on variation of maximum stress ($\sigma_{maximum}$) with fatigue life (N_F) for alloy steel 300M when cyclically deformed at room temperature (27°C) at a load ratio of 0.1**

To have a better understanding of the stress versus fatigue life (N_f) response of this high strength alloy steel in the two chosen orientations, longitudinal (L) and transverse (T), the variation of maximum stress with fatigue life (N_f) is replotted in terms of variation of maximum elastic strain [$\sigma_{max.}$ / E] to take into account the actual elastic modulus of the chosen high strength alloy steel along a given orientation as a function of fatigue life (N_f). This is shown in **Figure 4** and reveals a linear trend. A noticeable difference in behavior is observed at all values of maximum elastic strain [$\sigma_{max.}$ / E] for the two chosen orientations, i.e., longitudinal (L) and transverse (T). At equivalent values of maximum elastic strain the transverse orientation revealed higher fatigue life than the longitudinal counterpart. This clearly suggests the role of intrinsic microstructural features in governing the stress-fatigue life response as a function of test specimen orientation.

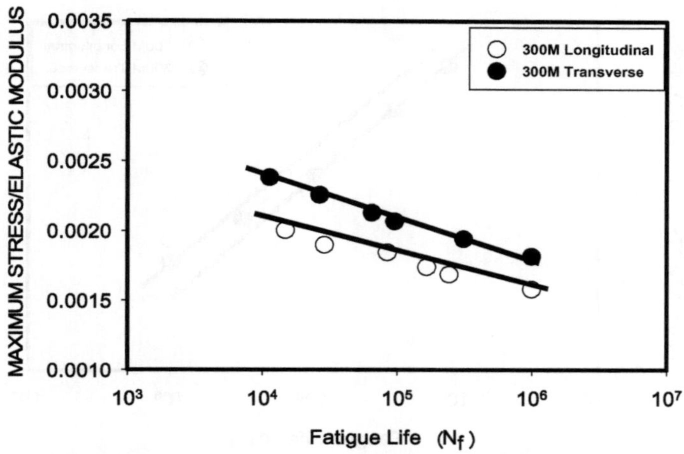

Figure 4. Influence of test specimen orientation on variation of maximum elastic strain ($\sigma_{max.}$ / E) with fatigue life (N_f) for alloy steel 300M when cyclically deformed at room temperature (27°C) and at a load ratio of 0.1.

To facilitate having a better understanding of the specific influence of strength as a function of orientation on high cycle fatigue response of the chosen high strength alloy steel, the variation of fatigue life (N_f) with maximum stress is replotted to highlight the specific influence of strength of this alloy steel. At a fixed ratio of maximum stress to yield stress [$\sigma_{max.}$ / $\sigma_{yield\ stress}$] the response of this alloy steel was marginally inferior in the transverse orientation when compared to the longitudinal orientation for test specimens that were cyclically deformed at a load ratio of 0.1. This is shown in **Figure 5**. The trend depicting the variation of maximum stress to ultimate tensile strength [$\sigma_{maximum}$ / σ_{UTS}] with fatigue life (N_f) is shown in **Figure 6** for samples that were cyclically deformed at a load ratio of 0.1. At equivalent values of this ratio the samples taken from the L orientation revealed observable reduction in cyclic fatigue life when compared to specimens taken from the T orientation. At a given value of this ratio the reduction in fatigue life is as high as 100-150 percent.

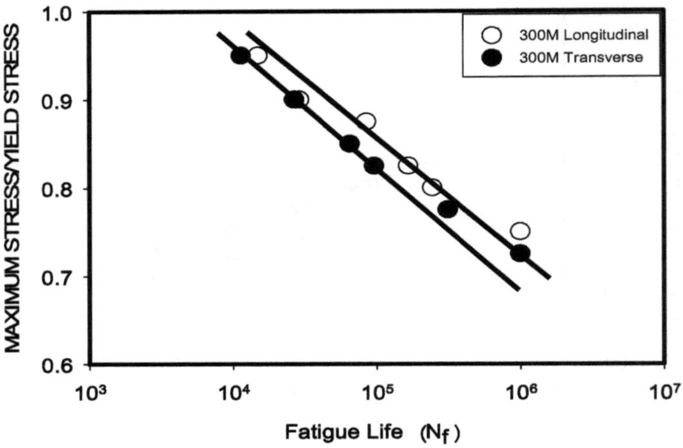

Figure 5. **Influence of test specimen orientation on variation of ratio of maximum stress to yield stress [$\sigma_{maximum}$ / σ_{YS}] with fatigue life (N_F) for alloy steel 300M cyclically deformed at room temperature (27°C).**

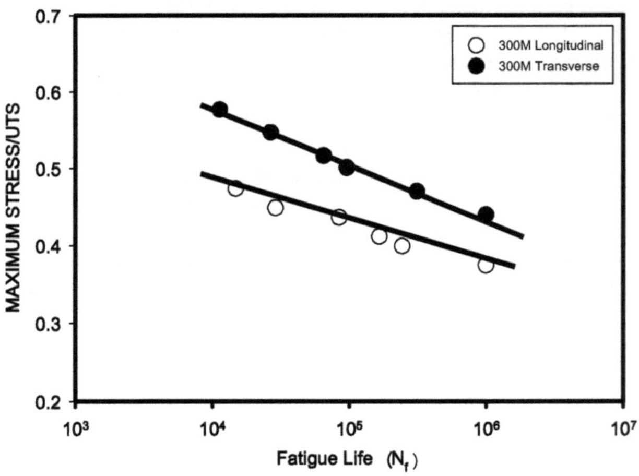

Figure 6: **Influence of test specimen orientation on variation of ratio of maximum stress to ultimate tensile strength [$\sigma_{max.}$ / σ_{UTS}] with fatigue life (N_f) for alloy steel 300 M cyclically deformed at room temperature (27°C).**

4.4 Cyclic Fracture Behavior

4.4.1 Longitudinal Orientation

A scanning electron micrograph of the fatigue fracture surface of the cyclically deformed and failed 300M test sample at a maximum stress of 425 MPa and having a resultant fatigue life of 14,897 cycles is shown in **Figure 7**. Overall morphology of failure at the macroscopic level was essentially normal, i.e., 90 degrees, to the far-field applied stress axis **(Figure 7a)** and microscopically very rough at the fine microscopic level (**Figure 7b**). High magnification observation of the region of fatigue and early microscopic crack growth revealed pockets of distinctly transgranular regions inlaid with an observable population of fine microscopic voids and isolated fine microscopic cracks **(Figure 7c)**. The region of unstable crack growth prior to overload revealed a smooth surface that was inlaid with easily observable fine microscopic cracks. The microscopic cracks were oriented in the direction of the major stress axis. Presence of this feature is indicative of the occurrence of "locally" operating brittle failure mechanism.

A sample of this alloy steel that was cyclically deformed at a lower maximum stress of 369 MPa and having a fatigue life of 166,459 cycles is shown in **Figure 8**. Overall morphology of failure was flat and essentially normal, i.e., 90 degrees, to the far field stress axis, i.e., tensile axis (**Figure 8a**). High magnification observation of the fatigue region revealed an array of fine microscopic cracks intermingled with observable macroscopic cracks (**Figure 8b**). High magnification observation of (b) shows the nature and orientation of the coplanar microscopic cracks present in the region of unstable crack growth prior to overload (**Figure 8c**). The region of overload revealed a noticeable population of fine microscopic voids of varying size intermingled with pockets of shallow dimples (**Figure 8d**). These features are indicative of the "locally" occurring ductile failure mechanisms.

4.4.2 Transverse Orientation.

Scanning electron micrographs showing overall morphology of failure and intrinsic features on the surface of a sample taken from the transverse orientation and cyclically deformed at a maximum stress of 490MPa giving a fatigue life of 26,737 cycles are shown in **Figure 9**. Overall morphology was both flat and at an angle to the far field stress axis (**Figure 9a**). Observation of this region and early microscopic crack growth at the higher allowable magnification of the SEM revealed an array of fine microscopic cracks and isolated macroscopic cracks traversing the grain boundaries (**Figure 9b).** The region of unstable crack growth prior to overload revealed an observable population of microscopic voids of varying size intermingled with shallow dimples **(Figure 9-c**). These features are indicative of the occurrence of "locally" ductile failure mechanisms. The region of fatigue overload revealed numerous fine microscopic voids intermingled with shallow dimples, features indicative of the improved ductility of this alloy steel in transverse orientation.

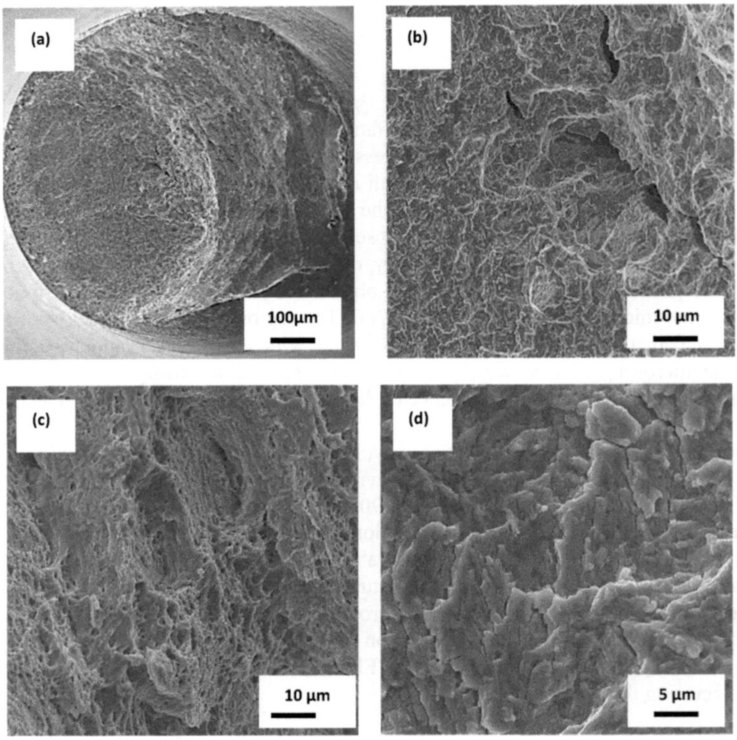

Figure 7. Scanning electron micrographs of the longitudinal test sample of 300 M cyclically deformed at a maximum stress of 425.MPa and a fatigue life of 14,897 cycles, showing:

(a) Overall morphology of failure normal to stress axis
(b) High magnification of the fatigue region showing microscopically rough fracture surface
(c) Fine microscopic voids and microscopic cracks in region of early microscopic crack.ne microscopic cracks
(d) Fine microscopic cracks in region of unstable crack growth.

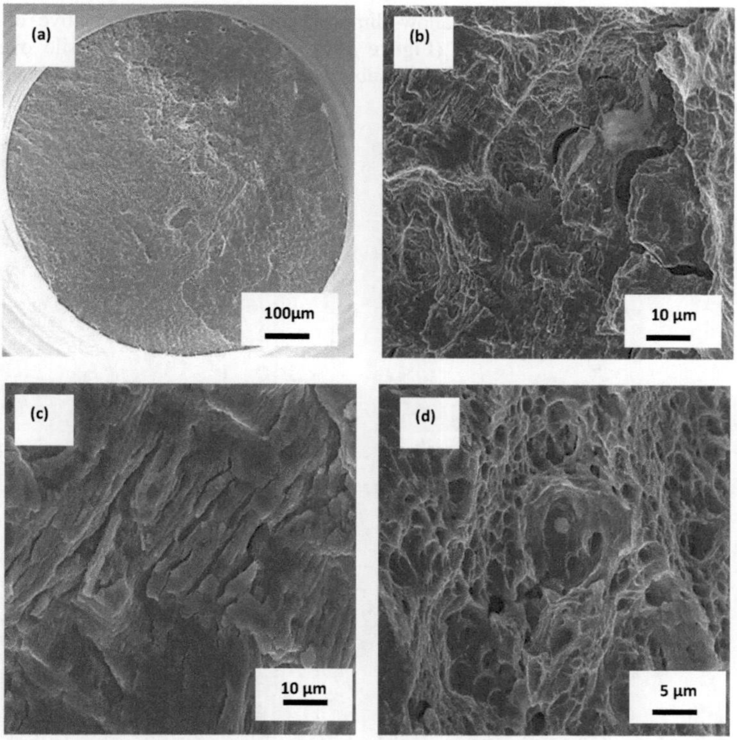

Figure 8. **Scanning electron micrographs of the longitudinal test sample of 300 M cyclically deformed at a maximum stress of 369 MPa and a fatigue life of 166,549 cycles, showing:**

(a) Overall morphology of failure

(b) High magnification of (a) showing macro-cracking in the fatigue region.

(c) Coplanar array of fine microscopic cracks in region of unstable crack growth..

(d) Population of shallow dimples intermingled with microscopic voids in the region of overload

A test sample of this high strength alloy steel taken from the transverse orientation and cyclically deformed at a maximum stress of 449 MPa with a fatigue life of 96,549 cycles is shown in **Figure 10**. Overall morphology of the fatigue fracture surface at the macroscopic level was at an inclination to the applied stress axis and essentially cup and cone morphology (**Figure 10a**). The region of fatigue and early microscopic crack growth when observed at higher magnification was flat, near featureless and primarily transgranular (**Figure 10b**). The region of unstable crack growth prior to overload revealed isolated fine microscopic voids interdispersed with an observable population of shallow dimples; features clearly indicative of "locally" operating ductile failure mechanisms **(Figure 10c)**. In the region of tensile overload the population of dimples was distinct and observable, indicative of the locally dominating ductile failure mechanisms.

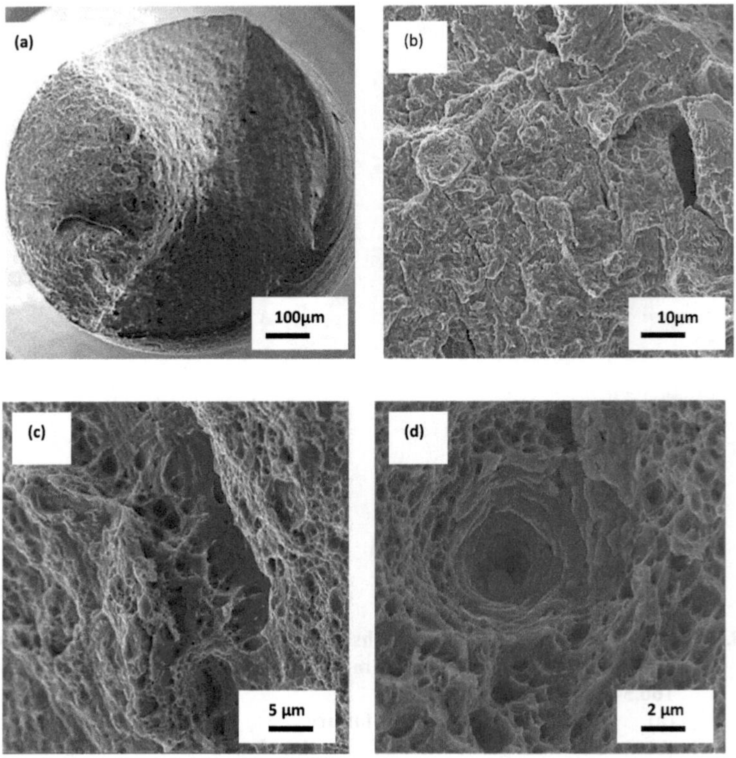

Figure 9. **Scanning electron micrographs of the transverse test sample of 300 M cyclically deformed at a maximum stress of 490 MPa and a fatigue life of 26,737 cycles, showing:**
(a) Overall morphology of failure
(b) Fine microscopic cracks and isolated macroscopic cracks in region of early crack growth.
(c) Population of microscopic voids and shallow dimples in region of unstable crack growth
(d) Fine microscopic voids and shallow dimples covering the overload fracture surface.

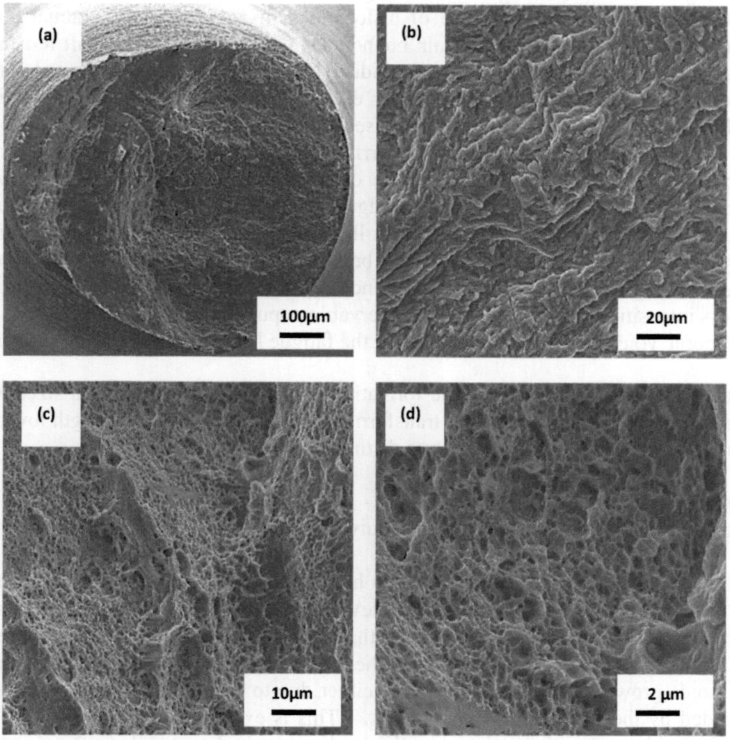

Figure 10. Scanning electron micrographs of the transverse test sample of 300 M cyclically deformed at a maximum stress of 449 MPa and a fatigue life of 96,549 cycles, showing:

(a) Overall cup and cone morphology of failure

(b) Early microscopic crack growth, flat, featureless and transgranular.

(c) Fine microscopic voids and shallow dimples in region of unstable crack growth.

(d) Population of dimples of varying size on the overload fracture surface.

4.5 Mechanisms Governing Stress-Orientation-Microstructure Interactions.

In this study an attempt was made at evaluating the specific influence of test specimen orientation on cyclic fatigue life (N_f) and resultant fracture behavior of the high strength alloy steel 300M. The observed fracture behavior is quantified on the basis of both the macroscopic mode and the key mechanisms occurring at the fine microscopic level.

During continuous loading in cyclic fatigue under conditions of tension-tension (positive load ratio of 0.1) loading the presence of dislocations, if any, and their gradual build up or accumulation at locations of: (a) the grain boundaries, (b) interfaces of the soft ferrite and hard martensite phases, and (c) other coarse carbide particles present and dispersed through the microstructure of this alloy steel, does aid in an early initiation of fine microscopic voids. This is favored to occur when the "local" strain caused by the progressive built-up of dislocations at interfaces of the (a) micro-constituents, i.e. ferrite and martensite, and (b) the matrix - second phase particle, reaches a critical value. The early nucleation of a fine microscopic void is favored to occur upon the initiation of yielding at the 'local' level. During continued cyclic loading a few of the hard and essentially elastically deforming second-phase particles are favored to fail by cracking on account of their intrinsic brittleness. Since both crack extension and void growth culminating in their eventual coalescence are favored to occur at the prevailing high 'local' stress intensities, the presence of an observable population of both macroscopic voids and fine microscopic voids contributes to degrading the fatigue life (N_f) of the test specimen.

At the fine microscopic level, the formation and physical presence of a sizeable number of fine microscopic voids contributes to transforming the deforming high strength low alloy steel into a composite material comprising of two distinct populations:

(i) Grains in the metal matrix, and
(ii) Voids [a void being considered to be equivalent to a particle having zero stiffness].

Since the voids are intrinsically softer than the hardened grains and other micro-constituents in the metal matrix, the "local" strain becomes elevated both at and around the region of the fine microscopic voids thereby enabling conditions that are conducive for an increase in their volume fraction. During continued cyclic loading the fine microscopic voids get elongated. The elongated voids grow and eventually coalesce either due to mutual interaction or impingement, which is aided by the formation of void-sheets. This is exemplified in **Figure 11.** Eventual failure of the void sheet with continued cyclic loading results in a rough fracture surface at the fine microscopic level. Half of the void is the shallow dimples that were observed distributed on the fracture surface.

5. Conclusions

The results of a study aimed at establishing an understanding of the specific influence of test specimen orientation on microstructure and their synergistic interactions on high cycle fatigue response, quantified in terms of life-to-failure, and resultant fracture behavior of the chosen high strength low alloy steel 300M provides the following findings:

1. The initial microstructure of the as-provided high strength steel revealed a combination of carbon-rich and carbon-depleted regions. The carbon-rich region was sizeable and noticeable. The presence of both carbon and alloy elements in this alloy steel resulted in substantial volume fraction of the micro-constituent martensite in the carbon-rich regions in comparison to the carbon-depleted region. The morphology of martensite was predominantly of the lath type.

2. The elastic modulus of the steel was between 212 GPa in the longitudinal orientation and 217 GPa in the transverse orientation. The yield strength of alloy steel 300 M was 14 pct. higher in the transverse orientation (544 MPa) than in the longitudinal orientation (477 MPa). The ultimate tensile strength of 300 M in the transverse orientation (922 MPa) was higher than the tensile strength in the longitudinal orientation (877 MPa). For both orientations the tensile strength is higher than the yield strength indicating the occurrence of strain hardening beyond yield. The ductility of this alloy steel was higher in the transverse orientation (17 pct.) than in the longitudinal orientation (15 pct.).

3. Variation of stress amplitude with fatigue life revealed a degradation in fatigue life for the longitudinal orientation under conditions of tension-tension loading. The degradation in cyclic fatigue life was more or less uniform over the entire range of maximum stress this alloy steel was cyclically deformed. When the chosen high strength alloy steel is examined on the basis of: (i) maximum elastic strain (σ_{max}/E), (ii) ratio of maximum stress to yield stress ($\sigma_{max}/\sigma_{yield}$), and (iii) maximum stress to ultimate tensile strength ($\sigma_{max}/\sigma_{UTS}$), the performance is observably inferior for the longitudinal orientation when compared to the transverse orientation.

4. The fatigue fracture surfaces of test samples of alloy steel, i.e., 300 M, revealed minimal difference in: (a) the overall morphology of failure, and (b) the nature, volume fraction and dispersion of fine microscopic features, as a function of maximum stress, at the chosen load ratio of $R = 0.1$.

5. For both orientations of alloy steel 300M, over the range of maximum stress the region of fatigue and early microscopic crack growth was essentially flat. This region was found to be flat, transgranular and lacking in discernible features. The region of unstable crack growth was short and revealed cracking spanning both microscopic and macroscopic cracks along with voids and dimples; features indicative of locally acting brittle and ductile mechanisms. The region of overload was sizeable for all of the samples examined in the microscope and contained an observable population of microscopic voids intermingled with isolated macroscopic voids and dimples of varying size and shape. These features are indicative of the locally occurring ductile failure mechanisms.

Acknowledgements

This research was made possible through material, samples and funding made possible by Latrobe Specialty Steel Company (Latrobe, PA, USA). ***Mr. Manigandan Kannan*** (Graduate Student: Department of Mechanical Engineering) extends thanks and appreciation to the University of Akron for providing support through tuition scholarship.

References

1. Innovations in Ultrahigh Strength Steel Technology, Proceedings of the 34th Sagamore Army Materials Conference (editors: G.B. Olson, M. Azrin and E. S. Wright), U.S. Army Materials Technology Laboratory, Watertown, MA, 1990, pp. 3-65.

2. G.B. Olson, M. Azrin, E.S. Wright: Innovations in Ultrahigh Strength Steel Technology (editors), U.S. Army Materials Technology Laboratory, Watertown, MA, 1990, pp. 89-109.

3. G. R., Speich, D.S. Dabkowski, L. F. Porter: **Metallurgical Transactions**, Vol., 4, 1973, pp. 303-315.

4. Thomas, G., Schmatz, D. and Gerberich, W.: "*Structure and Strength of Some Ausformed Steels", In 'High Strength materials*' (Editor. V.F. Zackey), John Wiley and Sons, New York, NY, 1965, pp. 251-326

5. Baker, A. J., Laura E. J., and Wei, R. P., "*Relationships Between Microstructure and Toughness in Quenched and Tempered Ultrahigh-Strength Steels*," Structure and Properties of Ultrahigh Strength Steels, STP 370, ASTM, Philadelphia, PA, 1965, pp. 3–30.

6. D. Huang and G. Thomas: "*Factors Influencing Strength and Toughness in High Strength Low Alloy Steels*," **Metallurgical Transactions**, 1972, Vol. 2, p. 1587

7. Honeycombe R.W.K, Harding H.J and Irani J.J.: "*Strengthening mechanism in Ferritic and Austenitic Steels* ", In 'High Strength Materials' (Editor: V.F. Zackey), John Wiley and Sons, New York, NY, 1965, pp. 213-250.

8. Baker, AJ.., Laura, F.D., and Wei, R.P., "*Structure and Properties of Ultra High Strength Steels"* ASTM STP 370, 1965, p 3.

9. G.E. Pellissier, "*Effects of microstructure on the fracture toughness of ultrahigh-strength steels*", **Engineering Fracture Mechanics**, Volume 1, Issue 1, June 1968, Pages 55-60.

10. Y. Tomita, **Materials Science and Technology**, Vol. 7, 1991m, pp. 481-490

11. Y. Tomitya and T. Okawa: " Effect of microstructure on mechanical properties of isothermally bainite transformed 300M steel, Materials Science and Engineering, Vol. 72, 1993, pp. 145-152.

12. E.W. Lee, C.E. Neu and J. Kozol: **Journal of Metals**, Vol. 42 (5), 1990, pp. 11-14.

13. D.Y. Wei, J.L. Gu, H.S. Fang, B.Z. Bai and Z.G. Yang: "*Fatigue behavior of 1500 MPa bainite/martensite duplex-phase high strength steel*", **International Journal of Fatigue**, Vol. 26, 2004, pp. 437-442.

14. Darius Rozumek: "*Influence of the slot inclination angle in $FeP0_4$ steel on fatigue crack growth under tension*", **Materials and Design**, Vol. 30, 2009, pp. 1859-1865.

15. Y. Furuya, S. Torizuka, E. Takeuchi, M. Bacher Hochst and M. Kuntz: "*Ultrasonic fatigue testing on notched and smooth specimens of ultrafine-grained steel*" **Materials and Design**, Vol. 37, 2012, pp. 515-520.

16. R. Hertzberg, Deformation and Fracture Mechanics of Engineering Materials, Second Edition, John Wiley and Sons, 1983,

17. D. Thomas, D. Schmatz and W.W. Gerberich: in High Strength Materials (Edited by: V.F. Zackay) John Wiley and Sons, New York, USA, 1965, pp. 199-209

18. W.M. Garrison, Jr., and N.R. Moody: "*The Influence of Inclusion Spacing and Microstructure on the Fracture Toughness of the Secondary Hardening Steel AF 1410*", **Metallurgical Transactions**, 1987, Vol. 18A, pp. 1257-1263.

Fatigue of Materials III: Advances and Emergences in Understanding
Edited by: T.S. Srivatsan, M. Ashraf Imam, and R. Srinivasan
TMS (The Minerals, Metals & Materials Society), 2014

THE EFFECTS OF PRE-STRAINING CONDITIONS ON AUSTENITE STABILITY DURING FATIGUE OF MULTIPHASE TRIP STEELS

Alexander L. Ly and **Kip O. Findley**

Colorado School of Mines
G.S. Ansell Department of Metallurgical & Materials Engineering
1500 Illinois St.; Golden
Colorado, CO 80401, **USA**

Abstract

A multiphase TRIP780 steel was subjected to tensile pre-strains of 0, 5, and 15% at room temperature and at -20 °C. The pre-strained specimens were subjected to strain-controlled, fully-reversed, axial fatigue testing at strain amplitudes ranging from 0.2 to 0.6%. The low-cycle fatigue life was largely independent of the pre-strain history. Pre-strained specimens had lower plastic strain amplitudes and higher stress ranges in the initial fatigue cycles before converging towards the same values as the specimens with zero pre-strain, due to cyclic strain softening. This suggests that the effects of pre-strain are eventually overcome by accumulated plastic deformation during fatigue cycling.

__Keywords__: TRIP steel, low-cycle fatigue, pre-strain, retained austenite

1. Introduction

Transformation-induced plasticity (TRIP) steels are of particular interest to the automotive industry due to their high strength and ductility [1-3]. The favorable combination of strength and ductility allow TRIP steels to absorb large amounts of energy before rupture occurs. The enhanced elongation achievable relative to other steels of comparable strength is attributed to a dispersion of retained austenite in the TRIP steel microstructure. As the steel is deformed, retained austenite transforms into martensite; this transformation results in increased strain-hardening, delaying the onset of localized plasticity [3]. In automobiles, TRIP steel properties allow for thinner sheets to be used without sacrificing performance, lowering the auto-body weight, reducing emissions, and increasing fuel economy [2-4].

Due to their excellent tensile and fatigue properties, TRIP steels are good candidates for underbody parts [2], which are often subjected to fatigue-loading conditions [4]; thus, the fatigue behavior of TRIP steels is highly relevant. Also, since sheet steel components must be formed before being put into service, it is important to consider the fatigue properties of TRIP steels after pre-straining [5, 6]. During these forming processes, heating occurs as a result of plastic deformation and friction (deformation-induced heating) [7, 8]; temperature increases of up to 100 °C are possible [7]. Aside from the rise in temperature from deformation-induced heating, temperature control may be used advantageously during forming; Deighton and Lee have shown that increases in forming limit are possible using punch temperatures as low as 5 °C [9]. Because temperature affects the propensity and mechanism of austenite-to-martensite transformation (stress or strain-induced) in TRIP steels, pre-strain temperature is expected to influence performance. Thus, it is relevant to investigate how both pre-strain level and pre-strain temperature influence TRIP steel behavior.

2. Experimental

The material to be investigated in this study is a bare, cold-rolled, multiphase TRIP780 steel with a nominal thickness of 2 mm and a chemical composition as displayed in Table I. Measured room-temperature tensile properties are shown in Table II. A field emission scanning electron micrograph of the TRIP steel is shown in Figure 1. The sample was mounted in Bakelite with the rolling plane exposed and polished to a 1-μm finish using diamond paste, then etched using a 2% nital solution (2% nitric acid, 98% ethanol by volume) to reveal the microstructure. In this micrograph, the sunken regions are ferrite, the raised globular regions are islands of austenite and martensite, and the needle-like regions are likely bainite.

Table I. Approximate Chemical Composition of Cold-Rolled TRIP780 Steel

weight percent	C	Mn	Si	Al	P	S	N
TRIP780	0.197	1.65	1.727	0.042	0.015	0.002	0.0034

Table II. Measured Tensile Properties of TRIP780 Steel

Yield Strength	495 MPa
Tensile Strength	854 MPa
Uniform Elongation	24.5%
Total Elongation	35.5%

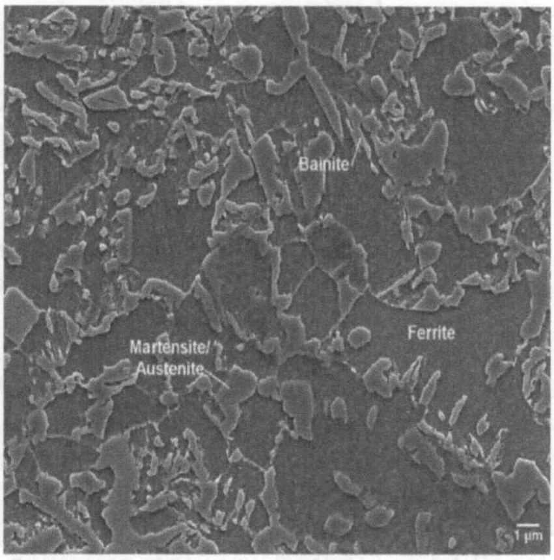

Figure 1. Field emission scanning electron micrograph of the as-received TRIP780 steel (rolling direction in the horizontal direction) etched with 2% nital. Ferrite, bainite, and islands of martensite/austenite are shown.

Figure 2 is a drawing of the fatigue specimen geometry. The as-machined specimens were chemically polished in a solution made of 7 parts water, 3 parts 30% hydrogen peroxide, and 1 part 48% hydrofluoric acid to produce a surface finish suitable for fatigue. This solution is placed in a water bath to keep the temperature constant and to mitigate heating of the specimen. After the vigorous reaction has calmed, the sample is removed from the acid bath and immediately rinsed with methanol. The sample is finally cleaned off using a polishing wheel with a 6-μm diamond suspension.

Prior to fatigue testing, a subset of specimens was pre-strained to 5 or 15% at room temperature or -20 °C. The temperatures were chosen based on $M_s^{\square}$, a quantitative parameter for austenite stability against stress-assisted transformation. Below the $M_s^{\square}$ temperature (and above the martensite start temperature below which martensite forms upon cooling), mechanical stress can be applied to nucleate martensite from retained austenite; this is known as stress-assisted nucleation. The required stress increases with temperature; when the required stress is equal to the austenite yield strength, the corresponding temperature is $M_s^{\square}$□□□Above $M_s^{\square}$, plastic deformation of austenite occurs and shear bands intersect, creating new nucleation sites for the phase transformation; this is strain-induced nucleation. The $M_s^{\square}$ temperature was determined to be approximately 10 °C; details of this measurement can be found in a previous report by the authors [10]. Thus, the pre-strain temperatures represent temperatures above (RT) and below (-20 °C) the $M_s^{□}$□temperature.

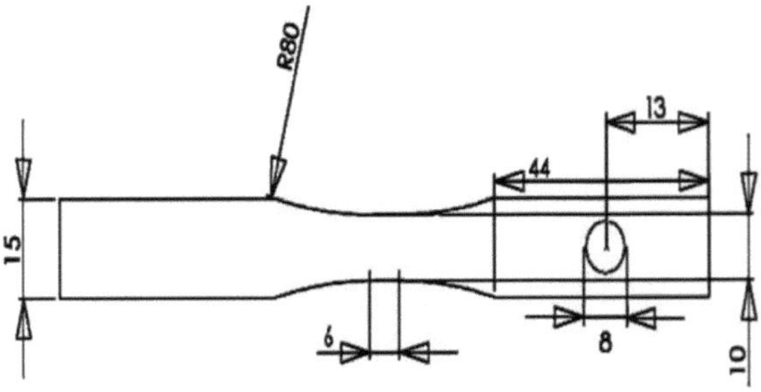

Figure 2. TRIP780 fatigue specimen with a hole in one grip for submersed pre-straining. All units are in mm, and the nominal thickness is 2 mm.

Fully-reversed, strain-controlled axial fatigue tests, with total strain amplitudes of 0.2, 0.3, 0.4, 0.5, or 0.6% were performed, using a triangular waveform with a strain rate of 0.032 per second. A 3-mm extensometer (Epsilon Tech) was attached to the side of the specimen to measure and control strain. Cyanoacrylate adhesive was applied to the specimen at the points of contact with the extensometer knife edges in order to reduce the chances of extensometer slipping and the creation of vulnerable sites for crack nucleation. To prevent buckling, an anti-buckling guide was used for all fatigue tests. For further protection against buckling at strain amplitudes of at least 0.4%, two specimens were glued together in a pack after pre-straining each separately. M-Bond 200 was also used to bond the specimens together.

X-ray diffraction (XRD) was performed to estimate the fraction of retained austenite present in the specimen at varying levels of pre-strain. To prepare a specimen for XRD analysis, a small sample was cut from the gauge section of a deformed tensile bar, ground down to a 600-grit finish, then placed into a chemical thinning solution (10 parts water, 10 parts 30% hydrogen peroxide, 1 part 48% hydrofluoric acid) for 15 minutes. The sample was then ultrasonically cleaned for five minutes. X-ray diffraction was performed at room temperature with copper radiation using a Phillips X-pert diffractometer operating at 45 kV and 40 mA. An X'celerator detector and 1-degree slit were used. The sample was scanned from 2-theta values between 40 and 105 with a step size of 0.008 2-theta. To estimate the austenite volume fraction, four γ peaks ({111}, {200}, {220}, and {311}) and four α peaks ({110}, {200}, {211}, and {220}) were used in the calculation.

3. Results and Discussion

X-ray diffraction measurements were performed on specimens strained at room temperature to degrees ranging from 0 to 23% in order to estimate the volume fraction of retained austenite as a function of strain. As Figure 3 shows, the amount of retained austenite present in the steel decreases with strain due to strain-induced martensitic transformation. In the as-received state, the TRIP780 steel contains approximately 16% retained austenite by volume.

After 5% strain, almost 50% of the austenite has transformed into martensite, and by approximately 15% strain, the austenite has been depleted. Thus, the three pre-strain levels used in the fatigue tests represent different amounts of prior martensitic transformation as well as different amounts of pre-strain.

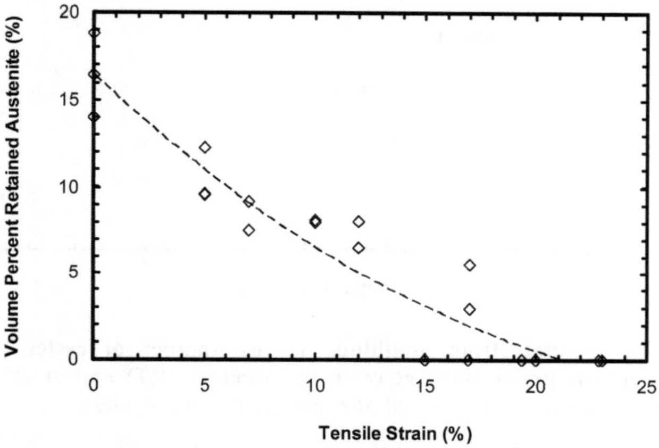

Figure 3. Plot of estimated retained austenite content based on XRD measurements as a function of tensile strain at room temperature.

Figure 4 plots total strain amplitude against the number of cycles to failure for specimens pre-strained at room temperature and -20 °C. This graph suggests that pre-straining has little to no effect on the low-cycle fatigue life of the steel within the inherent scatter of the fatigue results. A possible explanation for this similar behavior may be because crack nucleation dominates the fatigue life. As illustrated in Figure 1, the microstructure of this multiphase TRIP steel is mostly composed of ferrite, which is the softest phase. Low-cycle fatigue damage accumulation and crack nucleation is then most likely to occur in ferrite. Since ferrite is still the softest phase after pre-straining, the martensite formed during pre-straining may not have a direct effect on crack nucleation. By contrast, pre-straining does increase the fatigue life in the high-cycle fatigue region where the fatigue lives are controlled more by material strength, which increases with pre-straining, than material ductility. As Figure 4 shows, one non-strained specimen cycled at 0.2% total strain amplitude failed before reaching 1,000,000 cycles, while all pre-strained specimens did not fail before 1,000,000 cycles.

Although pre-straining may not have a significant effect on the fatigue life of the multiphase TRIP steel, it does affect the plastic strain amplitude behavior, as seen in the plastic strain amplitude versus cycle number plots in Figure 5. For a given total strain amplitude, increasing the level of pre-strain at room temperature decreases the initial plastic strain amplitude, since pre-straining raises the yield point and consequently the amount of elastic strain. The difference in plastic strain amplitude is less pronounced at lower total strain amplitudes because the fatigue deformation is already mostly elastic.

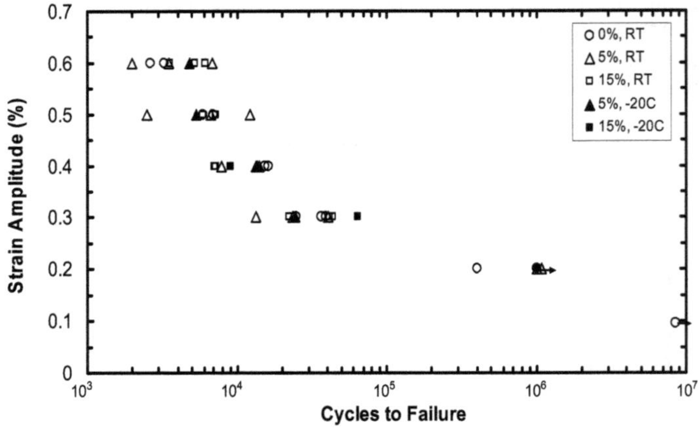

Figure 4. Plot of total strain amplitude versus number of cycles to failure for specimens pre-strained at room temperature (RT) and at -20 °C. Arrows denote fatigue tests stopped at runout (10^6 or 10^7 cycles).

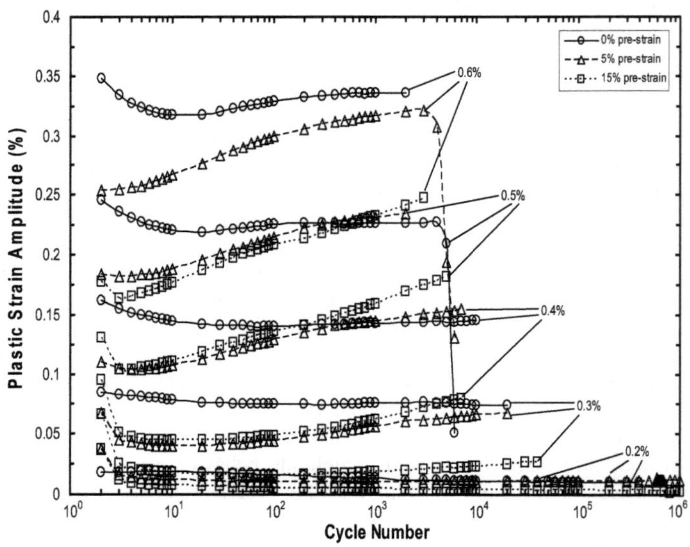

Figure 5. Plot of plastic strain amplitude versus cycle number for specimens pre-strained at room temperature. Labels denote total strain amplitude for each fatigue test.

Figure 5 also shows that there are consistent differences in the cyclic strain hardening and softening behaviors between each pre-strain condition. In the non-strained specimens, there is an initial decrease in plastic strain amplitude, indicating hardening; this initial hardening rate decreases with total strain amplitude. Higher strain amplitudes are expected to result in more martensite formation, which may account for increased hardening rates. At higher total strain amplitudes, this region is followed by cyclic softening and then saturation; at lower total strain amplitudes, hardening continues up until saturation. As martensite formation induces plasticity in the surrounding ferrite matrix, softening may be due to the resulting dislocation tangles rearranging themselves into more stable configurations. Pre-strained specimens produced plastic strain amplitude curves which were almost all monotonically increased during fatigue cycling, as dislocations produced during pre-straining were rearranged.

Interestingly, pre-straining seems to have negligible effect on the low-cycle fatigue life, despite decreasing the initial plastic strain amplitude. One possible reason is that the pre-strained conditions cyclically strain soften more than the non-strained condition. This is most apparent when comparing the 0% and 5% pre-strain cases for a given total strain amplitude in Figure 5. Plastic strain amplitude values may differ vastly in the early cycles, but they tend to converge as the test continues. The 0.4% total strain amplitude curves are isolated and displayed as a representative case in Figure 6 for clarity. However, it is apparent that the 15% pre-strain curves in Figures 5 and 6 do not quite converge with the 0% and 5% curves, though the gap separating the 15% pre-strain condition from the 0% and 5% pre-strain conditions decreases during fatigue cycling. Although the plastic strain amplitude is lower after 15% strain, the stress range is significantly higher, as shown in Figure 7. The higher imposed stress may counteract the effect of the lower plastic strain amplitude, resulting in a relatively unchanged fatigue life.

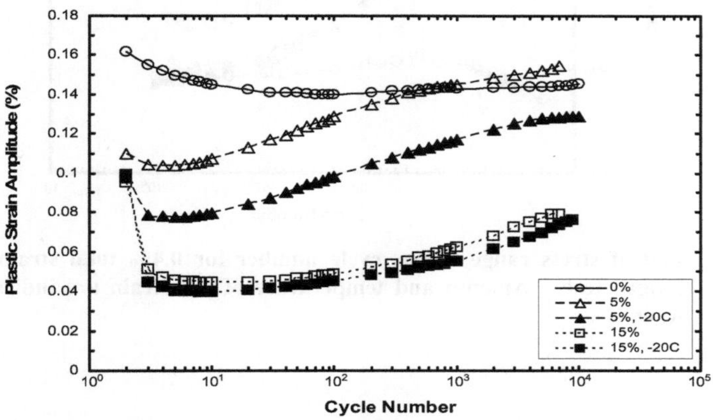

Figure 6. Plot of plastic strain amplitude versus cycle number for 0.4% total strain amplitude fatigue tests. Amount and temperature of pre-strain are indicated by the legend.

Two curves for specimens pre-strained at -20 °C are also included in Figure 6. Comparing the -20 °C curves to their room temperature counterparts reveals that lowering the pre-strain temperature seems to reduce the plastic strain amplitude but the curves retain the general shape as the specimens pre-strained to the same amount at room temperature. At reduced temperatures, stress-induced martensite formation from austenite becomes more favorable. Increased martensite content is consistent with the decrease in plastic strain amplitude, especially after the initial cycle. There is less difference between the 15% curves pre-strained at different temperatures than the 5% curves, since most of the retained austenite should have transformed even at room temperature when pre-strained to 15%; thus, pre-straining to 15% at lower temperatures did little to change the metallurgical structure that was ultimately fatigue tested. The resemblance between the shapes of the curves indicates that cyclic hardening and softening occur at similar rates after the initial hardening. In contrast, the cyclic hardening and softening behavior differs considerably after varying pre-strain levels and seems to be more dependent on degree of pre-strain than on temperature. This suggests that cyclic hardening and softening rates may be relatively independent of deformation-induced martensitic transformation. The softening behavior after the initial cycles is likely a result of dislocation network rearrangement in the ferrite and thus is mostly unaffected by the martensite content.

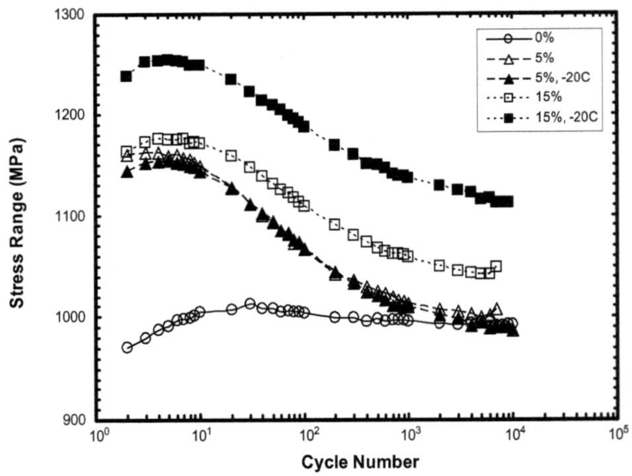

Figure 7. **Plot of stress range versus cycle number for 0.4% total strain amplitude fatigue tests. Amount and temperature of pre-strain are indicated by the legend.**

4. Summary and Conclusions

TRIP780 specimens were pre-strained below or above the $M_s^{\square}$ □temperature before performing low cycle fatigue tests. For the multiphase TRIP780 steel used in this study, the amount of retained austenite present ranged from approximately 16% with no prior straining to essentially zero after 15% tensile strain at room temperature. Pre-straining in tension decreases the initial plastic strain amplitude at a given total strain amplitude, as it increases the yield point. However, pre-straining does not have a significant effect on the fatigue life of this steel; plastic damage is likely concentrated in the soft ferrite phase, which is relatively unaffected by the deformation-induced martensite formation during pre-straining. Lowering the temperature of pre-strain decreased the plastic strain amplitudes while keeping similar cyclic hardening and softening rates as conditions pre-strained at room temperature. The increased martensite formation at reduced temperatures is likely responsible for the lower plastic strain amplitudes and higher cyclic hardening rates in the initial cycles, but hardening and softening behavior seem to be dictated mostly by the level of pre-strain rather than the martensite content, which is again probably due to deformation being concentrated in the ferrite during fatigue.

Acknowledgements

The authors would like to thank Dr. Shrikant Bhat of ArcelorMittal Steel for providing technical assistance and the TRIP780 steel used in this study. The Advanced Steel Processing and Products Research Center at the Colorado School of Mines is appreciated for its support and facilities. The National Science Foundation is also acknowledged for funding this project through grant DMR-0955236.

References

1. G. R. Chanani and S. D. Antolovich, "Low Cycle Fatigue of a High Strength Metastable Austenitic Steel," ***Metallurgical Transactions***, 5 (1974), 217-229.

2. L. Tosal-Martínez, D. Vanderschueren, S. Jacobs, and S. Vandeputte, "Development of a Hot-Rolled Nb-Bearing Si-TRIP Steel with Excellent Fatigue Behaviour for Automotive Applications," *Steel Research*, 72 (10) (2001), 412-415.

3. J. G. Speer and D. K. Matlock, "Recent Developments in Low-Carbon Sheet Steels," *JOM*, July 2002, 19-24.

4. Y. Sakuma, N. Kimura, A. Itami, S. Hiwatashi, O. Kawano, and K. Sakata, "Next-Generation High-Strength Sheet Steel Utilizing Transformation-Induced Plasticity (TRIP) Effect" (Report UDC669.14.018.2-41, Nippon Steel, Japan, March 1995).

5. L. T. Robertson, T. B. Hilditch, and P. D. Hodgson, "The Effect of Pre-Strain and Bake Hardening on the Low-Cycle Fatigue Properties of TRIP Steel," ***International Journal of Fatigue***, 30 (2008), 587-594.

6. S. M. Song, K. Sugimoto, S. Kandaka, A. Futamura, M. Kobayashi, and S. Masuda, "Effects of Pre-straining on High-Cycle Fatigue Strength of High-Strength Low Alloy TRIP-Aided Steels," ***Materials Science Research International***, 9 (3) (2003), 223-229.

7. M. Takahashi, "Development of High Strength Steels for Automobiles" (Report UDC699.14.018.295-415:629.11.011, Nippon Steel, Japan, July 2003).

8. Y. H. Kim and R. H. Wagoner, "An Analytical Investigation of Deformation-Induced Heating in Tensile Testing," ***International Journal of Mechanical Sciences***, 29 (3) (1987), 179-194.

9. D. E. Leighton and D. Lee, "The Effect of Tooling Temperature on the Formability of Sheet Steel," ***Journal of Materials Processing Technology***, 45 (1994), 577-582.

10. A. L. Ly and K. O. Findley, "The Effects of Pre-Straining Conditions on Austenite Stability during Fatigue of Multiphase TRIP Steels" (ASPPRC Research Report: Sheet and Coated Steels, MT-SRC-013-006, March 2013).

Fatigue of Materials III: Advances and Emergences in Understanding
Edited by: T.S. Srivatsan, M. Ashraf Imam, and R. Srinivasan
TMS (The Minerals, Metals & Materials Society), 2014

FATIGUE CHARACTERISTIC OF MEDIUM CARBON STEEL WITH DIFFERENT GRAIN STRUCTURES

Sulaiman Abdulkareem[1], A.T. Abdulrahim[2], S. I. Talabi[3] and O. H. Amuda[4]

[1] *Department of Mechanical Engineering,*
Materials Development and Characterization Research Group
University of Ilorin, P.M.B. 1515, Ilorin, **Nigeria**
E-Mail: sulkarm@yahoo.com

[2] *Department of Mechanical Engineering,*
University of Maiduguri, P.M.B 1069
Maiduguri, Borno State, **Nigeria**
E-Mail: engrabdulrahimat@gmail.com

[3] *Department of Materials and Metallurgical Engineering,*
Materials Development and Characterization Research Group
University of Ilorin, P.M.B. 1515
Ilorin, **Nigeria**
E-Mail: isaacton@yahoo.com

[4]Department of Metallurgical and Materials Engineering
University of Lagos, Nigeria
E-Mail: chatwivwale@yahoo.com

Abstract

This paper investigated the effect of heat treatment operations on the fatigue resistance of low carbon steel. Specimens after preparation for fatigue testing were subjected to annealing, normalizing and quenching heat treatment. Results show that the annealed specimen had the largest number of cycles to failure, indicating a high fatigue resistance. The microstructure of the specimens was examined in other to corroborate the obtained property with the microstructure. When compared with the untreated specimen, the annealed specimen (with optimum fatigue resistance) shows a large grains size of pearlite which was distributed across the entire surface of the microstructure. Generally, it was found that the size and distribution of specimens' grains affect the resistance of the low carbon steel to fatigue failure.

Keywords: fatigue, medium carbon steel, heat treatment, microstructure

1. Introduction

Majority of engineering designs, structures, machines, and systems involve materials analysis, selection and categorization. These engineering materials are numerous and among them are metals, composites, ceramics and plastics. The most useful one being metals due to their mechanical properties like tensile and compressive strength, ductility, brittleness, fatigue strength and others [1]. Among these metals, low carbon steel has a very high status in terms of its versatility, cheapness, ease and low cost of production which makes it the most widely used metallic material. The mechanical properties (and some other properties like electrical, chemical) form the major characterization and categorization of metals and also determines the type of material needed for certain specific engineering purposes. Hence due to the suitability of metals for very wide range of applications, the more need to study and establish certain facts regarding metals arise.

Also, some properties of the mild steel might be required to be varied due to design specifications and requirements, which is mostly done by various heat treatment operations [2]. These heat treatments give the desired properties of carbon steel, but sometimes at the expense of some other properties. Also, heat treatments results come about only by the systematic alteration of established micro structural properties of the materials [3]. In other words, heat treatment alters the micro structural arrangements and sizes of the metals grains which are the reasons for the changes observed in mechanical properties of metals after heat treatment operations.

Hence, before selecting a type of heat treatment, various factors need to be considered including the type of loading the materials will be subjected to in the service conditions. Of the various loading types is fatigue loading which is a load whose magnitude varies with (or is a function of) time. Hence, using fatigue loading as a case study, it would be observed that different materials fail at different load cycles and this character can be directly traced to the different grain structures, sizes and orientation of the materials [4].

Metallic materials consist of a microstructure of small crystals called grains or crystallites. The grain sizes and composition which make up the nature of the grains are one of the most effective factors that determine the overall mechanical behaviors (hardness, tensile strength, toughness, wear resistance) of metals [5, 7]. An efficient way to manipulate the properties of metals via grain structure manipulation by controlling the rate of diffusion and that of cooling is that of heat treating procedures.

2. Materials and Methods

The chemical composition of the as-received specimen is presented in Table 1. The specimens were prepared for fatigue test by machining on a lathe machine. The specimens were then subjected to annealing, normalising and quenching. In the annealing operation specimen were heated to an austenizing temperature of 850°C±5°C, soaked for 20 minutes in order to attain uniform heat distribution and left to cool in the muffle furnace (Power – 5kW, Phase – 1, Weight – 55kg, Temperature – 1200°C, Voltage – 220V, Dimension of chamber – 300x200x120mm). Similar operation was carried out during normalizing and quenching operation except that the specimens after heating were cooled in air and quenched in silica sand respectively. Fatigue test was done with is an Avery Denison type with serial no 7300. Grips are provided for torsion tests and for bend tests on flat specimen. Standard microstructural test pieces (as-received and heat treated) were prepared and ground using emery paper with grit 220 to 600 microns in succession. The ground surface was polished using a mixture of alumina and diamond paste and then etched in a solution containing 2 ml Nitric acid and 100 ml Ethanol. The etched surfaces were left for 20 seconds before they were rinsed and dried. The specimens' crystals morphology was viewed under an optical metallographic microscope at 600X magnification and the photomicrographs are shown in Figures1-4.

Table 1. Composition of Specimen

Elements	% wt Composition
Iron	99.030
Carbon	0.1164
Silicon	0.0901
Manganese	0.3574
Phosphorous	0.0297
Chromium	0.0231
Nickel	0.0767
Copper	0.1178
Vanadium	0.0010

3. Results and Discussions

3.1 Effect of Heat Treatment on the Microstructure of Specimens

Figure 1 shows a uniform distribution of ferrite which are the brightly coloured parts and pearlite (a mixture of ferrite and cementite) which are the dark or black spots and which are congested along the grain boundaries and also slightly dispersed over the ferrite. The grain sizes are moderate and the boundaries appear smooth, apart from the pearlitic presence.

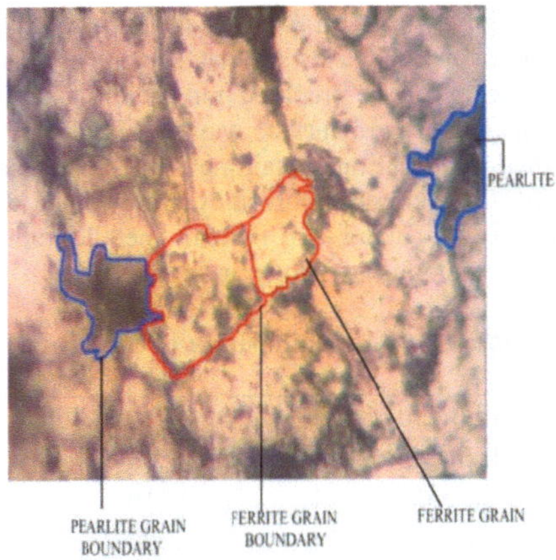

Figure 1. Micrograph of as-received specimen at X600

Figure 2 is the annealed specimen and the micrograph shows a large grain size of pearlite which was distributed across the microstructure. Unlike the untreated specimen, the pearlite phase which was congested along the grain boundaries is now distributed across the entire surface of the microstructure. The size of the grains makes it easy for dislocation to glide during fatigue loading, hence the high fatigue resistance. Annealed materials are usually ductile and stress free due to these stated characteristics. In addition, when a crack or void forms in a pearlitic matrix, it will tend to run along the length of a pearlite lamella. Examining thistype of fracture under the SEM reveals that the base of the dimples contains fractured pearlite lamella [2].

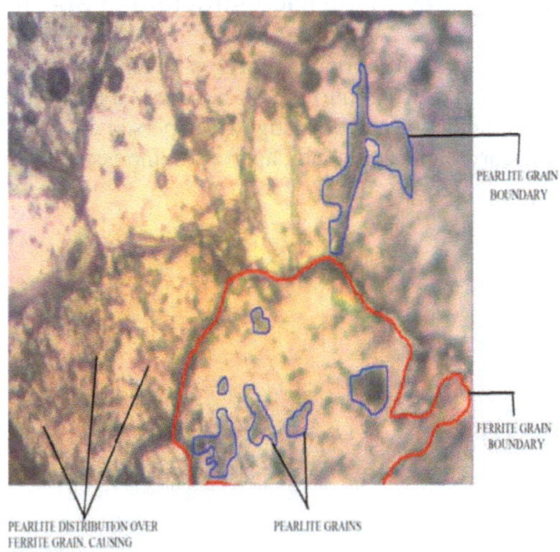

Figure 2. Micrograph of Annealed Specimen at X600

When compared to the annealed sample, the micrograph of the normalized sample (Figure 3) shows a relatively smaller grain size and a balanced distribution of ferrite and pearlite with uniform fine grain and quite similar to the control specimen, but appears relatively better than annealed and control samples

In addition, the sand cooled specimen shows a fine and relatively smaller ferrite and pearlite distribution and has a relatively lower amount of pearlite concentration as observed in the control and annealed specimen and it is observed that the grain boundaries appear thin and congested (Figure 4).

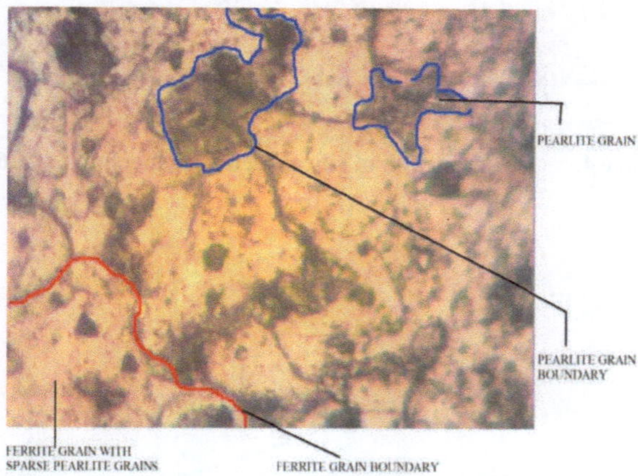

Figure 3. Micrograph of normalised specimen at X600

3.2 Effect of Heat Treatment on the Fatigue Property of Specimens

As seen from the results from graphs (Figures 5-8), a significant variation in the no. of cycles to failure was observed with respect to all the tested specimens. The annealed specimen had the largest number of cycles to failure, indicating a high fatigue resistance, followed by the specimen that was quenched in silica sand and then the normalized specimen. This agrees with the microstructures of the annealed specimen has it exhibit large and coarse distribution of pearlite [4].

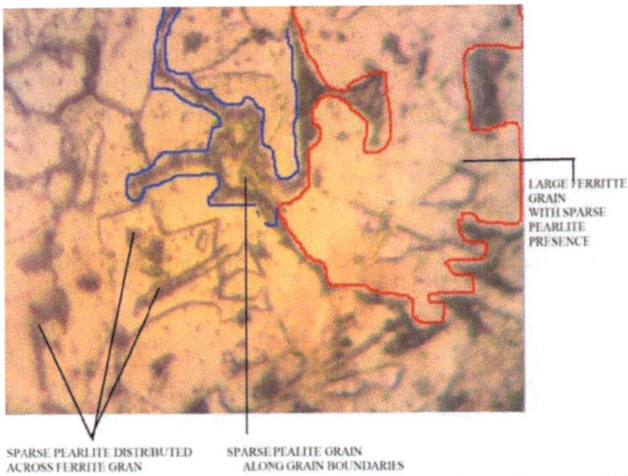

Figure 4. Micrograph of silica quenched specimen at X600

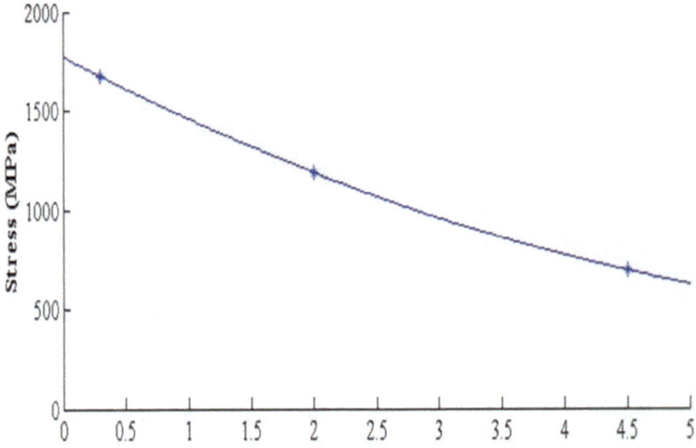

Figure 5. Stress-Fatigue life curve for the as-received specimen

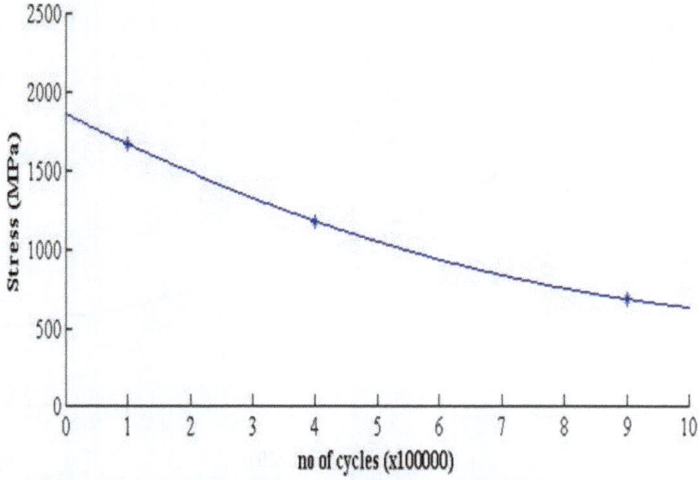

Figure 6. **Stress-Fatigue life curve for the annealed specimen**

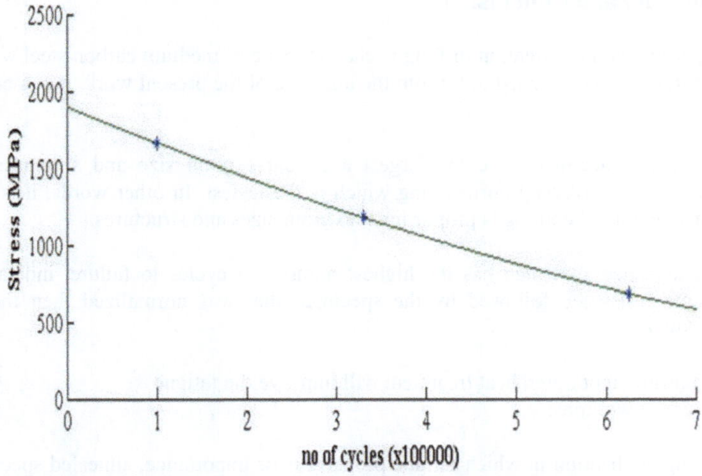

Figure 7. **Stress-Fatigue Life curve for the normalized specimen**

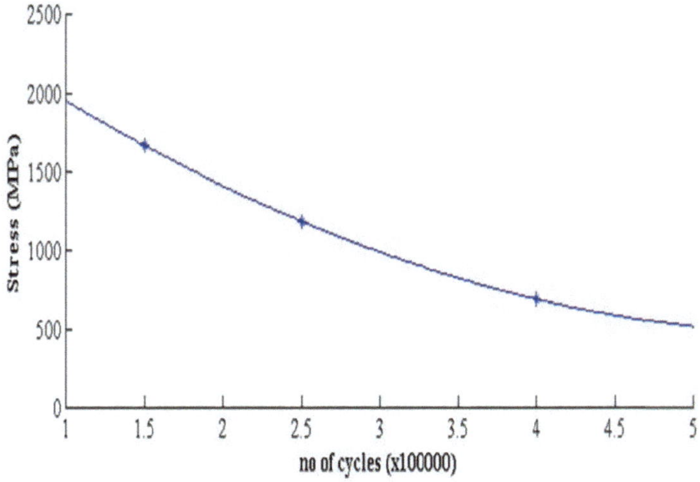

Figure 8. **Stress-Fatigue Life curve for the quenched specimen**

4. Summary and Conclusions

Experimental investigation of fatigue characteristic of medium carbon steel with different grain structures has been carried out. From the outcome of the present work, it can be concluded that:

1. Annealing specimen gave the largest and coarse grain size and structure, while the smallest was given by normalizing which is the fastest. In other words, it is established that slow rate of cooling is proportional to grain sizes and structures.

2. The annealed specimen has the highest number of cycles to failure, indicating a high fatigue resistance followed by the specimen that was normalized then the quenched workpiece.

3. A carefully monitored heat treatment will improve the fatigue life of the workpiece.

4. During application in which fatigue property is of importance, annealed specimen which shows a better fatigue resistance is hereby recommended.

References

1. D. J. Davis and L. A. Oelmann *"The structures, properties and Heat Treatment of metals"*, The Pitman Press, Great Britain, 1983, pp. 131-153.

2. Scott Mackenzie, *"Overview of the Mechanisms of Failure in Heat Treated Steel Components"*, Failure Analysis of Heat Treated Steel Component, ASM International, Materials Park, Ohio, USA, 2008, pp. 1-45.

3. G. E. Totten (Editor), "Steel Heat Treatment Handbook". vol. 1. *Metallurgy and Technologies*, Equipment and Process Design, 2nd edition, CRC Press, Boca Raton, Florida, USA, 2007.

4. T. Senthilkumar, and T. K. Ajiboye, *"Effect of Heat Treatment Processes on the Mechanical Properties of Medium Carbon Steel."* Journal of Minerals & Materials Characterization & Engineering, Vol. 11, 2012, pp. 143-152.

5. S. Z. Qamar: *"Effect of Heat Treatment on Mechanical properties of H11 tool steel,"* . Journal of Achiv Mat and Manufacturing Engineering, Vol. 35, 2009, pp. 115–120.

6. Heat Treatment of Steel: Application and Heat Treatment of Carbon and Alloy Steels, Machinery's Hand Book, Society of Automobile Engineer, 1974.

7. Adhan Calik, *"Effect of Cooling Rate on Hardness and Microstructure of AISI 1020, AISI 1040 and AISI 1060 Steels"*, Internatiional Journal of Physical Sciences, Vol. 4, 2009, pp. 514-527.

FATIGUE of MATERIALS III
ADVANCES and EMERGENCES in UNDERSTANDING

Modeling

ON THE ROLE OF PLASTICITY-INDUCED FATIGUE CRACK CLOSURE IN HIGH-STRENGTH STEELS

J. Toribio, V. Kharin

Fracture of Materials and Structural Integrity Research Group
University of Salamanca, Spain
E-Mail: toribio@usal.es, vikharin@gmail.com

Abstract

Plasticity-induced crack closure (PICC) has long been focused as supposedly controlling factor of fatigue crack growth (FCG). However, when the plane-strain near-tip constraint is approached, PICC lacks of straightforward evidence, so that its significance in FCG, and even the very existence, remain debatable. To add insights into this matter, large-deformation elastoplastic simulations of plane-strain crack under constant amplitude load cycling at different load ranges and ratios, as well as with an overload, have been performed. Modeling visualizes the Laird-Smith conceptual mechanism of FCG by plastic blunting and re-sharpening. Simulation reproduces the experimental trends of FCG concerning the roles of stress intensity factor range and overload, but PICC has never been detected.

Keywords: Crack closure, Crack-tip strains, Fatigue cracking

1. Introduction

The capability of opposite crack surfaces to contact prematurely at unloading due to particular patterns of the near-tip plastic straining under cyclic loading, the so called plasticity induced crack closure (PICC), is believed to be the intrinsic feature of fatigue crack growth (FCG), cf., e.g., the Overview [1], although not everybody involved in fatigue studies shares this conviction [2-4]. Despite the phenomenon of crack closure has been focused for decades since its raising by Elber [5], neither agreement between evaluations by different ways nor consensus about the relevance, and even the very existence, of PICC in the course of FCG do exist, cf. [1-3,6,7]. Conclusions about the PICC mechanisms based on both dislocation and continuum plasticity analyses are contradictory, cf., e.g., Deshpande *et al.* [8] or Pippan and Riemelmoser [9] *vs.* Louat *et al.* [3] or Bjerkén and Melin [10], Budianski and Hutchinson [11] *vs.* Noroozi *et al.* [12], and McClung *et al.* [13] *vs.* Toribio and Kharin [14].

Crack closure seems to be out of doubts as a phenomenon that can accompany FCG and affect it under certain circumstances. Reviews [1,6,7] catalogued a variety of ways of the crack closure verification and assessment ranging from deductions based on the crack growth behavior to interpreting the compliance measurements [5]. Thus the identification of PICC lacks of evidence and its responsibilities in FCG remain debatable.

Since *in situ* visualization of the near-tip deformations is hardly feasible in the places where the plane-strain situation is approached, computational simulation turns out to be the right way to determine them. Accounting for both constitutive (inelasticity) and geometrical (large displacements and strains) nonlinearities is here essential. For cracks growing by means of local material rupture, the alteration of solid boundary via crack advance by bond breaking is the third nonlinearity.

Up to date, among available analyses of fatigue cracks, some of them have not accounted for large deformations (such as, e.g., [13,15,16]), whereas others (e.g., [14,17-24]), although fulfilling this deficiency, presented partial data about cracks under cyclic loading. Anywise, common shortcoming is the lack of realistic treatment of the crack growth by means of bond breaking or node release to the analyst's discretion at the bottom, mid or top points of designated loading cycles irrespectively of both the material and applied load (cf., e.g., [13,16,21,24]), and thus, having hardly whatsoever to do with the physics of crack advance.

The efforts involving more realistic mechanisms of material damage and bond breaking directly in the modeling of FCG have been undertaken since nearly a decade (e.g., [25,26]) usually by substantial computational expenses. However, these relatively few attempts are at rather early stage of development and encounter conceptual and technical difficulties on the way to implement damage mechanics in large-deformation formulations. Another approach counts on the stress-, strain- or energy-based variables as fatigue damage monitors [27-30].

Meanwhile, among conceptual mechanisms of FCG, the one, which was suggested by Laird and Smith [31] and reaffirmed by Pelloux [32] and Neumann [33], and which was emphasized as the rational physical model for FCG in ductile materials [2,10,22,23,34,35], relies solely on the near tip plastic deformations under cyclic loading without involvement of pull-apart bond breaking. Then, high-resolution large-deformation elastoplastic simulation is the right way to visualize this mode of FCG. Recent advances in large-strain elastoplastic analyses of cracks under cyclic loading [14,17-20,22,23] revealed various aspects of FCG according to the Laird-Smith concept.

The scope of this paper is to narrate the results of modeling of cracks under cyclic loading, which are in particular association with the matter of PICC verification and assessment, and have not been revealed in previous presentations [14,18-20] of the performed simulations. Deformations near the crack subjected to mode I cyclic loading under plane strain and small scale yielding (SSY) are described with the intention to bring insight into some long-standing controversies about the crack blunting—re-sharpening and PICC under load cycling.

2. Modelling

The results presented herein proceed from the extensive modeling of the mode I crack tip fields in elastoplastic material under SSY in terms of the crack tip autonomy dominated solely by the stress intensity factor (SIF) K [36]. Model design (geometry, loading and constitutive material model) and the large-strain analysis procedure were basically the same as presented and substantiated elsewhere [14,18-20].

At large strains, material hardening approaches saturation, so that elastic—perfectly-plastic constitutive model can be an acceptable approximation. The simulations were performed for constant amplitude loadings at different SIF ranges $\Delta K = K_{max} - K_{min}$, and ratios $R = K_{min}/K_{max}$, and the effect of a single overload was considered, along the following load cases:

(I) $\Delta K = 2K_0$, $R = 0$,
(II) $\Delta K = K_0$, $R = 0$,
(III) $\Delta K = K_0$, $R = 0.5$,
(IV) $\Delta K = K_0$, $R = 0$, with an overload to $K_{ov} = 2K_0$ at a single cycle of the loading path,
where the reference value $K_0 = 30$ MPa√m corresponds to loading regimes of fatigue cracking in steels [2,34].

The model of undeformed crack was parallel-flanks slot of the width b_0 with semicircular tip, Fig. 1(*a*), as it has been repeatedly substantiated and used in analyses of cracks, e.g., by McMeeking [37]; Needleman and Tvergaard [38], Rice *et al.* [39],] or Toribio and Kharin [14,18-20]. The value of $b_0 = 5$ μm was taken as a reasonable choice for steels [40]. Model design was subdued to ensuring the SSY in terms of the K-dominated crack tip autonomy [36]. The choice was made [19,20] in favor of full-scale specimens, which were double-edge-cracked panel (DECP) and central-cracked panel (CCP) under remote traction by uniform stress σ_{ap}, Fig. 1(*b*). Due to specimen symmetries, the boundary-value problems were stated for the same region $-a \le X \le W - a$, $0 \le Y \le H$ (Fig. 1(*b*), shadowed), being a the initial crack size, $2W$ and $2H$ the panel width and height, respectively. Boundary conditions were posed according to the specimen symmetries and loading, Fig. 1(*c*). To ensure the SSY, the model sizes were $a = 15000b_0$, $a/W = 0.2$ and $W = H$. The SIF calibrations were specified using the compendium [41] for DECP and CCP, respectively. The textbook [34,36] estimation for the monotonic plastic zone dimension under plane strain, $R_Y = K^2/(3\pi\sigma_Y^2)$, renders for the chosen material, geometry and loading $\max\{R_Y/a, R_Y/H, R_Y/W\} \le 0.014 << 1$, ensuring the SSY.

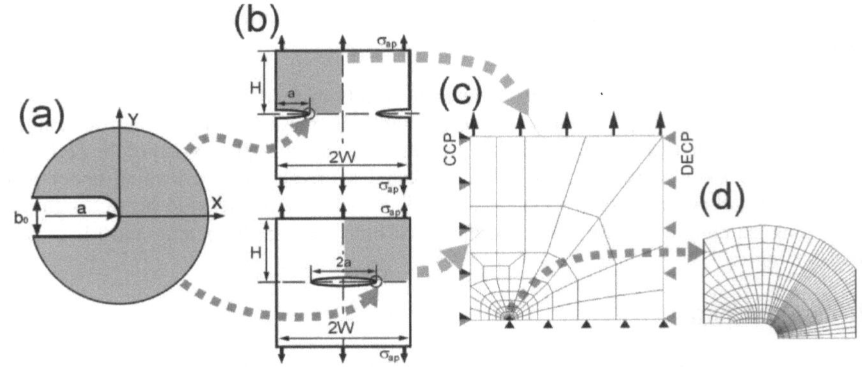

Figure 1: Model geometry and boundary conditions: (*a*) crack, (*b*) DECP and CCP specimens, (*c*) global view of the FE model of the specimen quarters shadowed in (*b*) with respective boundary conditions for CCP and DECP test-pieces, (d) near tip mesh refinement.

Large-deformation elastoplastic solutions were generated using a finite element (FE) code with updated lagrangian formulation and additive decomposition of strain rates. The FE mesh design followed previous studies of large near tip deformations [37-39]. To overcome numerical problems associated with mesh degeneration at the crack tip, extensive trials were accomplished to ensure the solution mesh convergence for acceptable number of loading cycles. Re-meshing was considered to be not a suitable means to improve the FE model performance because of substantiated doubts [42] that, under the propensity to solution bifurcation of certain elastoplasticity problems due to their loss of ellipticity [43,44. To this end, it must not surprise to meet severe discrepancies between near tip deformations obtained using different mesh layouts, FE formulations and re-meshing procedures, which rendered crack faces folding without closure at the very first load cycles [17], as well as smooth surfaces arriving at closure at fairly large number of cycles [22,23].

A variety of were explored attending the roles of FE size and shape, as well as of the interpolation order and integration scheme, as described elsewhere [42]. To postpone mesh degeneration, FE models of full-integration bi-linear quadrilateral elements were elaborated using rather high grade of element refinement.

3. Modelling

For all geometry-and-loading cases, the crack-tip deformations evolved as shown in Fig. 2. Plastic crack growth Δa_p is there appreciated as an advancement of the tip apex A_0 deeper into material. This way, the Laird-Smith conceptual scheme, which is considered to be one of the intrinsic mechanisms of FCG in ductile materials [2,31-35], is visualized. Simulations evidence that *the crack does grow* under cyclic loading, although with *no bond breaking*.

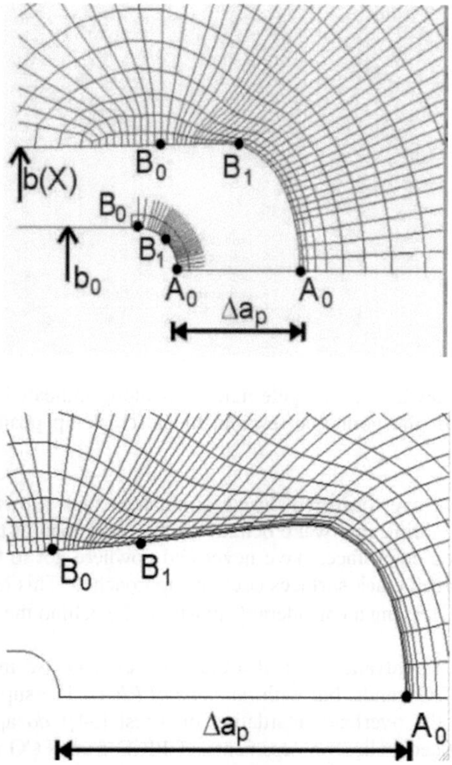

Figure 2: Crack tip deformations at unloading after accomplishing the first (*a*) and fifth (*b*) cycles of the load case I (DECP specimen); undeformed mesh fragment and tip contours are shown near the image center in (*a*) and in the bottom-left corner in (*b*), respectively, together with the positions of material points A_0, B_0 and B_1 in the initial and deformed configurations.

The plastic crack advancements Δa_p *vs.* cycle number N are presented in Fig. 3. The slopes of these curves render corresponding rates of plastic crack growth $(da/dN)_p$. Its acceleration with ΔK is there evident, whereas the effect of R is rather vague. Calculated $(da/dN)_p$ values are of the order of 10^{-6} m/cycle, which is proper for the Paris regime in steels [2,34]. The Paris-like equation $(da/dN)_p = C\,\Delta K^{m}$ fits the numerical results at $m \approx 2.25$, which is reasonable for many alloys [34]. This mode of crack extension is sensible to an overload, which halts the crack advance (Fig. 3, load case IV). This way, the effects of ΔK and overload on FCG obtained by simulations agree with common experimental trends [2,34].

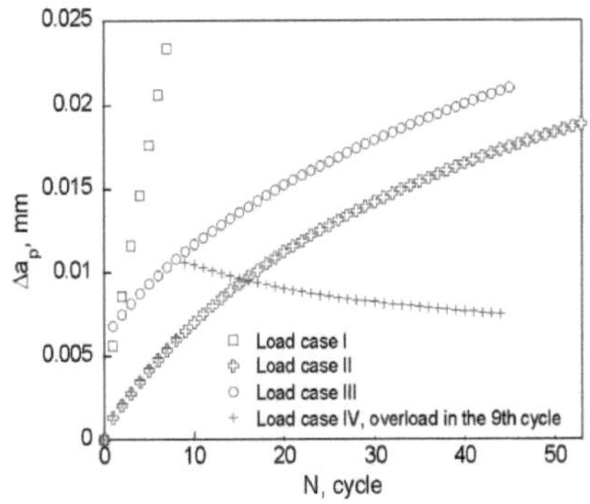

Figure 3. Crack growth □a_p *vs.* cycle number N along indicated loading routes in CCP specimen (the points correspond to the crack tip positions at the load cycle ends).

No signs of PICC have ever occurred during simulated crack growth at $R \geq 0$. Although cracks upon unloading shrink in a wake behind the tip more than at the tip itself in all load cases (Fig. 2), deformed crack faces have never and nowhere got to the initial crack width, and so, no contact between crack surfaces could be approached. This holds for an appreciable crack advancement Δa_p leaving a considerable plastic wake behind the tip.

Indeed, the plastic advancement of a crack behaves *per se* in agreement with well-known experimental FCG trends, but with *no signs of PICC*. The supposed consequences of closure in FCG (e.g., the overload retardation or arrest [34]) *do* appear with no closure. Accordingly, the practiced indirect assessments of PICC from FCG behavior [7] acquire a deal of spuriousness, which can make them artifacts.

Displayed results fairly agree with other large-strain analyses of cracks under cyclic loading [22,23], but challenge those small- and finite-strain simulations [13,16,21,24], where crack growth was imposed by means of cutting the material arbitrarily with a "virtual knife" in analyst's hands counting on neither the material nor the applied load. So, all these latter models were intrinsically unable to render *per se* any effect of the loading route on FCG, such as the influence of ΔK or overload. In contrast, presented modeling does this.

Moreover, by gluing FE meshes to the material, it is revealed here how the crack grows by plastic straining, which renders material transfer from the crack front onto its lateral surfaces, as it does a neighborhood of the material point B_1 in Fig. 2. Material "bricks" situated initially a little bit aside the top A_0 of the arc $B_0B_1A_0$ move there sideways and build the crack flanks up, i.e., contribute to crack enlargement. This is lengthened additionally by stretching of these material “bricks”, which are placed to form the crack flanks, Fig. 4(*a*), where large tensile plastic strain $\varepsilon_{yy}^{p} > 0$ in the in-plane of the crack direction is observed, Fig. 4(*b*), implying negative axial strain $\varepsilon_{YY} \approx -\varepsilon_{XX} < 0$ near the crack tip by virtue of plane-strain incompressible plasticity. This way, performed large-deformation simulations discard the mechanism of PICC implied from the small-strain modeling [13], where material elements behind the crack tip underwent out-of-plane plastic stretching $\varepsilon_{yy}^{p} > 0$ accompanied by shrinking along the crack surfaces $\varepsilon_{xx}^{p} \approx -\varepsilon_{yy}^{p} < 0$ by virtue of the plane-strain incompressible plasticity, Fig. 4(*c*). Accordingly, it was argued that this out-of-plane stretched material on the growing crack surfaces behind the tip could fill-in the crack, and this way enable the opposite crack surfaces to contact prematurely at unloading, i.e., produce PICC. This contradicts Figs. 2 and 4(*a,b*), where large displacements and rotations in the crack tip vicinity make material not to fill-in the crack behind the tip and render PICC, but to increment the crack flanks, i.e., they make the crack advance. This supports the consideration of the suggested PICC origin as an artifact [13], as far as it appears to be not the naturally present feature of a crack under cyclic loading but a product of the analysis method.

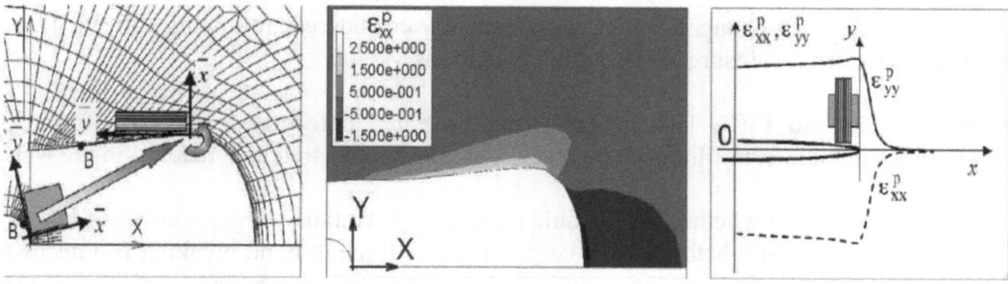

Figure 4. Deformations near the crack tip under cyclic loading:

(*a*) crack-tip large-deformation patterns with a scheme showing how and where do material "bricks" with their glued local material frames $(\bar{x}, \bar{y})$ go by translation and rotation (arrows) from the initial configuration to make the crack grow;

(*b*) contour bands of the plastic strain ε^p_{xx} at the end of the sixth cycle of the route I in deformed solid configuration (undeformed crack tip is seen in the bottom-left corner) to illustrate the contribution of the "bricks" stretching to crack lengthening, but not to its closure;

(*c*) scheme of the small-deformation results used to substantiate the suggested origin of PICC in the supposed filling-in the crack with stretched material behind the tip; the squares shaded in gray and rectangles filled with line pattern represent in (*a*) and (*c*) the same material elements in the initial and deformed states, respectively.

4. Conclusions

The crack becomes larger after every cycle, i.e., it grows, as suggested the Laird-Smith concept, by means of plastic straining with certain rate $(da/dN)_p$.

This growth reproduces the key experimental trends of FCG, such as the acceleration with ΔK and the hindering by overload, which have been repeatedly attributed to the crack closure.

Specimen compliance changes, which are frequently considered, following Elber [5], to be the evidences of crack closure, do occur in simulations.

Despite all that, no PICC has appeared, and revealed deformation patterns discard its supposed origination from filling-in the crack with plastically stretched material in the wake.

It does not matter now whether PICC could occur at substantially larger number of cycles or longer crack growth, or whether local material fracture through bond breaking is a necessary requisite for PICC to arise, but the point is that the supposed consequences of PICC do take place with no closure behind them, thereby bringing doubts about PICC.

Acknowledgements

The authors wish to acknowledge the financial support provided by the following Spanish Institutions: Ministry for Science and Technology (MCYT; Grant MAT2002-01831), Ministry for Education and Science (MEC; Grant BIA2005-08965), Ministry for Science and Innovation (MICINN; Grants BIA2008-06810, and BIA2011-27870) and *Junta de Castilla y León* (JCyL; Grants SA067A05, SA111A07, and SA039A08).

References

1. "Overview", *Advances in Fatigue Crack Closure Measurement and Analysis, ASTM STP 1343,* R.C. McClung, J.C. Newman, Eds., ASTM International, West Gonshohocken, 1999, p. xi.

2. Ritchie, R. O., "Mechanisms of crack propagation in ductile and brittle solids," Int. J. of Fract.,Vol. 100, 1999, pp. 55-83.

3. Louat, N., Sadananda, K., Duesbery, M. and Vasudevan, A. K., "A Theoretical Evaluation of Crack Closure," Met. Trans., Vol. A24, 1993, pp. 2225-2232.

4. Vasudevan, A. K., Sadananda, K. and Glinka, G., "Critical Parameters for Fatigue Damage," Int. J. of Fatigue, Vol. 23, 2001, pp. S39-S53.

5. Elber, W., "Fatigue Crack Growth under Cyclic Tension," Eng. Fract. Mech., Vol. 2, 1970, pp. 37-45.

6. Macha, D. E., Corbly, D. M. and Jones, J. W., "On the Variation of Fatigue-Crack-Opening Load with Measurement Location," Exp. Mech., Vol. 19, 1979, pp. 207-213.

7. Xu Yigeng, Gregson, P. J. and Sinclair, I., "Systematic Assessment of Compliance-Based Crack Closure Measurements in Fatigue," Mater. Sci. and Eng., Vol. A284, 2000, pp. 114-125.

8. Deshpande, V. S., Needleman, A. and van der Giessen, E., "A Discrete Dislocation Analysis of Near-Threshold Fatigue Crack Growth," Acta Mater., Vol. 49, 2001, pp. 3189-3203.

9. Pippan, R., Riemelmoser, F. O., "Visualization of the Plasticity-Induced Crack Closure under Plane Strain Conditions," Eng. Fract. Mech.,Vol. 60, 1998, pp. 315-322.

10. Bjerkén, C., Melin, S., "Growth of a Short Fatigue Crack – A Long Term Simulation Using a Dislocation Technique," Int. J. of Solids and Struct.," Vol. 46, 2009, pp. 1196-1204.

11. Budiansky, B., Hutchinson, J. W., "Analysis of Closure in Fatigue Crack Growth," J. of Appl. Mech., Vol. 45, 1978, pp. 267-276.

12. Noroozi, A. H., Glinka, G. and Lambert, S., "Prediction of fatigue Crack Growth under Constant Amplitude Loading and a Single Overload Based on Elasto-Plastic Crack Tip Stresses and Strains," Eng. Fract. Mech., Vol. 75, 2008, pp. 188-206.

13. McClung, R. C., Thacker, B. H. and Roy, S., "Finite Element Visualisation of Fatigue Crack Closure in Plane Stress and Plane Strain," Int. J. of Fract., Vol. 50, 1991, pp. 27-49.

14. Toribio, J., Kharin, V., "Large Crack-Tip Deformations and Plastic Crack Advance During Fatigue," Mater. Lett., Vol. 61, 2007, pp. 964-967.

15. Ellyin, F., Wu, J., "Elastic-Plastic Analysis of a Stationary Crack Under Cyclic Loading and Effect of Overload," Int. J. of Fract., Vol. 56, 1992, pp. 189-208.

16. Wu, J., Ellyin, F., "A Study of Fatigue Crack Closure by Elastic-Plastic Finite Element for Constant-Amplitude Loading," Int. J. of Fract., Vol. 82, 1996. pp, 43-65.

17. Levkovitch, V., Sievert, R. and Svendsen, B., "Simulation of Fatigue Crack Propagation in Ductile Metals by Blunting and Re-sharpening," Int. J. of Fract., Vol. 136, 2005, pp. 207-220.

18. Toribio, J., Kharin, V., 1998, "High-Resolution Numerical Modelling of Stress-Strain Fields in the Vicinity of a Crack Tip Subjected to Fatigue," Fracture from Defects, EMAS, 1998, pp. 1059-1064.

19. Toribio, J., Kharin, V., "Role of Fatigue Crack Closure Stresses in Hydrogen Assisted Cracking," *Advances in Fatigue Crack Closure Measurement and Analysis, ASTM STP 1343,* R.C. McClung, J.C. Newman, Eds., ASTM International, West Gonshohocken, 1999. p. 440.

20. Toribio, J., Kharin, V., "Finite Deformation Analysis of the Crack-Tip Fields under Cyclic Loading," Int. J. of Solids and Struct., Vol. 46, 2009, pp. 1937-1952.

21. Roychowdhury, S., Dodds, R. H., "A Numerical Investigation of 3-D Small-Scale Yielding Fatigue Crack Growth," Eng. Fract. Mech., Vol. 70, 2003, pp. 2363-2383.

22. Tvergaard, V., "On Fatigue Crack Growth in Ductile Materials by Crack-Tip Blunting," J. Mech. and Phys. of Solids, Vol. 52, 2004, pp. 2149-2166.

23. Tvergaard, V., "Overload Effects in Fatigue Crack Growth by Crack-Tip Blunting," Int. J. of Fatigue, Vol. 27, 2005, pp. 1389-1397.

24. Lei, Y., "Finite Element Crack Closure Analysis of a Compact Tension Specimen," Int. J. of Fatigue, Vol. 30, 2008, pp. 21-31.

25. Lynn, A. K., DuQuesnay, D. L., "Computer Simulation of Variable Amplitude Fatigue Crack Initiation Behaviour Using a New Strain-Based Cumulative Damage Model," Int. J. of Fatigue, Vol. 24, 2002, pp. 977-986.

26. Nguyen, O., Repetto, E., Ortiz, M. and Radovitzky, R., "A Cohesive Model of Fatigue Crack Growth," Int. J. of Fract., Vol. 110, 2001, pp. 351-369.

27. Chalant, G., Remy, L., "Model of Fatigue Crack Propagation by Damage Accumulation at the Crack Tip," Eng. Fract. Mech., Vol. 18, 1983, pp. 939-952.

28. Chan, K. S., Lankford, J., "A Crack Tip Strain Model for the Growth of Small Fatigue Cracks," Scripta Met., Vol. 17, 1983, pp. 529-532.

29. Fan, F., Kalnaus, S. and Jiang, Y., "Modelling of Fatigue Crack Growth of Stainless Steel 304L," Mech. of Mater., Vol. 40, 2008, pp. 961-973.

30. Hurtley, P. J., Evans, W. J., "A New Method for Predicting Fatigue Crack Propagation Rates," Mater. Sci. and Eng., Vol. A466, 2007, pp. 265-273.

31. Laird, C., Smith, G. C., "Crack Propagation in High Stress Fatigue," Phil. Mag., Vol. 8, 1962, pp. 847-857.

32. Pelloux, R. M. N., "Crack Extension by Alternating Shear," Eng. Fract. Mech., Vol. 1, 1970, pp. 697-704.

33. Neumann, P., "New Experiments Concerning the Slip Process at Propagating Fatigue Cracks," Acta Met., Vol. 22, 1974, pp. 1155-1165.

34. Suresh S., *Fatigue of Materials*, Cambridge University Press, Cambridge, 1991. p. 617.

35. Riemelmoser, F. O., Pippan, R. and Stüwe, H. P., "An argument for a cycle-by-cycle propagation of fatigue cracks at small stress intensity ranges," Acta Mater. Vol. 46, 1998, pp. 1793-1799.

36. Kanninen, M.F., Popelar, C.H., *Advanced Fracture Mechanics*, Oxford University Press, New York, 1985. p. 562.

37. McMeeking, R. M., "Finite Deformation Analysis of Crack Tip Opening in Elastic-Plastic Materials and Implications for Fracture," J. Mech. and Phys. of Solids, Vol 25, 1977, pp. 357-381.

38. Needleman, A., Tvergaard, V., "Crack-Tip Stress and Deformation Fields in a Solid With Vertex on its Yield Surface," *Elastic-Plastic Fracture: Second Symposium Vol. 1 — Inelastic Crack Analysis, ASTM STP 803,* C.F. Shih, J.P. Gudas, Eds., ASTM International, 1983, p. 80.

39. Rice, J. R., McMeeking, R. M, Parks, D. M. and Sorensen, E. P., "Recent Finite Element Studies in Plasticity and Fracture Mechanics" Comput. Meth. Applied Mech. and Eng., Vol. 17/18, 1979, pp. 411-442.

40. Handerhan, K. J., Garrison, W. M., Jr., "A Study of Crack Tip Blunting and The Influence of Blunting Behavior on the Fracture Toughness of Ultra High Strength Steels," Acta Met. et Mater., Vol. 40, 1992, pp. 1337-1355.

41. Savruk, M. P., *Stress Intensity Factors in Solids With Cracks*, Naukova Dumka, Kiev, 1988. p. 619

42. Toribio, J., Kharin, V., "Comments on Simulations of Fatigue Crack Propagation by Blunting and Re-sharpening: The Mesh Sensitivity," Int. J. of Fract., Vol. 140, 2006, pp. 285-292.

43. Hill, R., "Acceleration waves in solids," J. Mech. and Phys. of Solids, Vol. 10, 1962, pp. 1-16.

44. Rice, J. R., 1977, "The localization of plastic deformation," Theoretical and Applied Mechanics, North-Holland, Amsterdam, 1977, p. 207-220.

Fatigue of Materials III: Advances and Emergences in Understanding
Edited by: T.S. Srivatsan, M. Ashraf Imam, and R. Srinivasan
TMS (The Minerals, Metals & Materials Society), 2014

FATIGUE LIFE PREDICTION IN ALUMINUM ALLOY 2618-T6 USING A PARIS LAW MODIFICATION.

Salas-Zamarripa, A.[1], Pinna, C.[2], Brown, M.W.[2], Guerrero-Mata, M. P.[1], Castillo Morales, M.[1] and Ledezma-Ramírez, D.F.[1]

[1]Facultad de Ingeniería Mecánica y Eléctrica
Universidad Autónoma de Nuevo León.
Ave. Universidad S/N
Ciudad Universitaria, C.P. 66451, Apartado Postal 076
Suc. "F" San Nicolás de los Garza, N.L., **México.**

[2]Department of Mechanical Engineering
University of Sheffield.
Sir Frederick Mappin Building
Mappin Street, Sheffield, S1 3JD, **UK.**

Abstract

Fatigue tests at both room and elevated temperature (230°C) were carried out to understand the fatigue behavior of the 2618-T6 under such conditions. SEM fractography was performed in the fracture surfaces and a considerable amount of ductile intergranular fracture was found. A methodology based on stereophotogrammetry was developed to quantify the percentage of these areas on the fracture surfaces. The effect of these areas on the fatigue life was explored using a model based on a modification of the Paris law. The main idea of this model was to consider the material as the mixture of two phases associated with the two modes of failure observed on the fracture surfaces. The model gave a good prediction of the fatigue life measured during this research. Results from the model showed that the fatigue life is overestimated if the presence of the ductile intergranular fracture is not accounted for.

Keywords: Fatigue, Ductile intergranular fracture, Paris Law

1. Introduction.

In many precipitation hardening alloys, there are a condition in which the interior of the grains is highly hardened by precipitates while a section near grain boundaries remains softer because of the absence of precipitates known as Precipitate-Free Zones (PFZ), as shown schematically in Figure 1. PFZ can act as an ideal site for initiation and propagation of cracks, because plastic deformation can be confined inside those soft zones. Ductile intergranular fracture is related to the presence of these microstructural heterogeneities. This fracture occurs when microvoids formed around precipitates located along the grain boundary coalesce with voids growing inside the PFZ. The need to consider the PFZ as an independent microstructural feature has been pointed out by several authors [1-6].

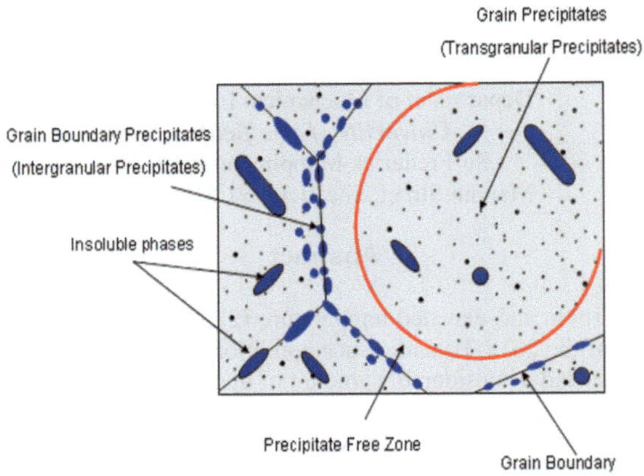

Figure 1. Representation of PFZ in the microstructure.

Ductile intergranular fracture was identified during inspection by scanning electron microscopy (SEM) in the fracture surfaces at both testing temperatures, as shown in Figure 2. This contribution was quantified by a method developed during this research using stereo-photogrammetry, and it is fully described in [7]. It was found in the fracture surfaces at room temperature, that the ductile intergranular fracture gradually reaches a maximum percentage as the crack growth and drops as extents to the final fracture. Meanwhile, fracture surfaces obtained from tests carried out at 230°C show a multiple-crack phenomenon (2 or 3 predominant cracks), where the ductile intergranular fracture contribution was less than at room temperature.

Several theories used to predict the fatigue life of a component are very complex. The use of Linear Elastic Fracture Mechanics (LEFM) concepts is still extremely popular in engineering practices as a result of the relative simplicity of its formulation. The Paris equation relates the fatigue crack growth rate (*da/dN*) to the stress intensity factor range (*ΔK*) as follows [8, 9]:

$$\frac{da}{dN} = C(\Delta K)^m \qquad (1)$$

where, *C* and *m* are materials constants.

However, the misuse of these LEFM concepts and neglect of significant parameters in the fatigue life calculations is common. It has been argued that the life prediction methodology must be customized for the specific application and not the other way around which is frequently the case [10].

a)

b)

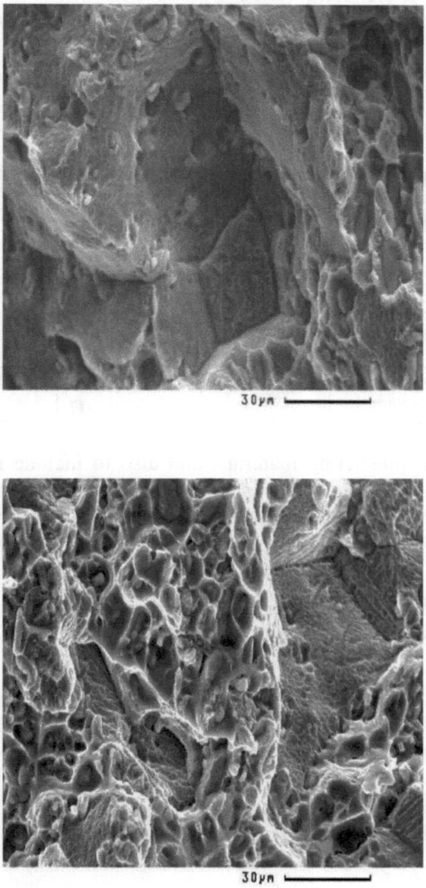

Figure 2. Ductile intergranular fracture: a) room temperature and b) 230°C.

A model to predict the fatigue life involving the effect of ductile intergranular fracture was developed during this research. The model is a modification of the Paris law. The model considers the 2618-T6 aluminium alloy as a two-phase material. The softer phase was modelled as pure aluminium (1100 aluminium alloy) and would represent the PFZ in order to quantify the ductile intergranular fracture contribution to the total fatigue life of the two-phase material. Results taken from the literature [11, 12] were used to develop this model.

2. Theoretical and experimental basis of the model.

It is known that the Paris parameters, *C* and *m*, are material parameters. The model developed in this research implied that the material behaves as two interacting materials; one described the ductile intergranular fracture contribution; and the other the rest of the fracture surface. This interaction is reflected in the Paris parameters. The mechanical behaviour of the ductile intergranular fracture contribution was described by a softer material, an 1100 aluminium alloy which is a pure aluminium alloy. This alloy has lower mechanical properties than the 2618-T6 aluminium alloy, as can be seen in Table 1. The rest of the fracture surface was characterized using the original mechanical properties of the material.

Table 1. Mechanical properties of materials used in the model developed [12-14].

Mechanical Properties	Aluminium alloy	
	1100	**2618-T6**
Yield strength (MPa)	**132**	**369**
Ultimate tensile strength (MPa)	**142**	**409**
Modulus of elasticity (GPa)	**68**	**74**
Hardness	**67-71 Rockwell H**	**80 Rockwell B**

The relation of the two interacting materials and also to include the ductile intergranular fracture contribution in the Paris parameters was based in the rule of mixtures for heterogonous materials *C* and *m* values of the two-phase material were calculated involving the percentages of ductile intergranular fracture measured experimentally [7], as:

$$C_T = (DI_F \times C_{1100}) + [(1 - DI_F) \times C_{2618}] \quad (2)$$

$$m_T = (DI_F \times m_{1100}) + [(1 - DI_F) \times m_{2618}] \quad (3)$$

where, suffix *T* refers to the total value and the suffixes 1100 and 2618-T6 correspond to the 1100 and 2618-T6 aluminium alloy respectively. DI_F is the ductile fracture contribution and is a function of crack length (*a*).

As mentioned before, *C* and *m* values were calculated from previous works and are listed in Table 2 [11, 12]. The fatigue tests performed during this research were carried out a load ratio, R, equal to 0.16. This ratio was considered to calculate the parameters values.

Table 2. Paris law parameters.

Parameters	*R*	*Aluminium alloy*	
		2618-T6	**1100**
C	**0.16**	**1.29E-08**	**6.11E-07**
	-1	**5.42E-09**	**3.63E-07**
m		**4.12**	**2.82**

Two approximations were used to include the ductile intergranular fracture contribution calculated from the fatigue fracture surfaces obtained at room temperature. One approximation was using all the data to obtain an average linear approximation of the experimental results (Figure 3); while the second one used the linear tendencies of each test, an example is shown in Figure 4.

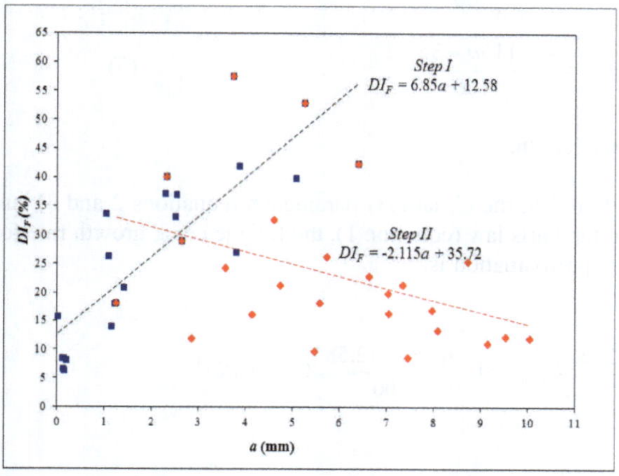

Figure 3. Ductile intergranular fracture contributions of the entire tests carried out at room temperature. The curves were divided in two stages: Initiation-*DI*$_F$ peak (squares) and *DI*$_F$ peak to failure (diamonds). The equations used are also indicated.

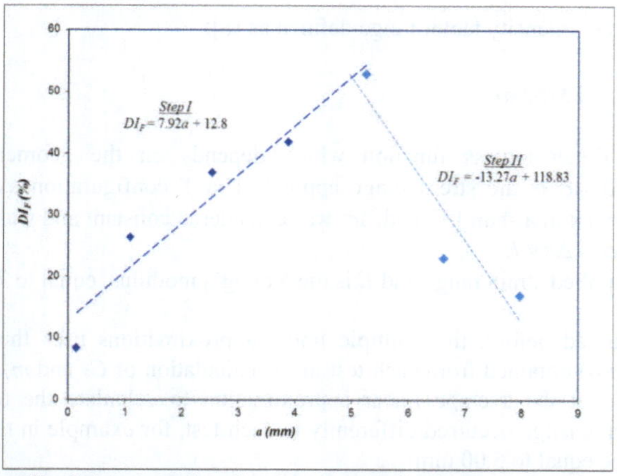

Figure 4. Ductile intergranular fracture contribution curve of the $\Delta\varepsilon/2 = 0.29\%$ indicated the linear tendency used in this research to run the multiple models.

The average linear approximation considered two steps. Step I includes all points between initiation and ductile intergranular fracture peak (indicated as squares in Figure 3), and step II consists in the measurements between ductile intergranular fracture peak and failure (shown as diamonds in Figure 3) of each test. The step change occurred when the two trend lines intersect for a given crack length (a). In this case a will be equal to 2.58 mm. For this approximation two equations were defined,

Step I $$DI_F = \frac{6.85a + 12.58}{100} \quad (4)$$

Step II $$DI_F = \frac{-2.115a + 35.72}{100} \quad (5)$$

where a is the crack length.

Substituting equation 4 in the C_T and m_T parameters (equations 2 and 3), and then including these equations in the Paris law (equation 1), the fatigue crack growth rate for the step I using the average linear approximation is,

$$\frac{da}{dN} = \left[\left(\frac{6.85a + 12.58}{100}\right)C_{1100} + \left(1 - \frac{6.85a + 12.58}{100}\right)C_{2618}\right](\Delta K)^{m_T} \quad (6)$$

accordingly step II will give,

$$\frac{da}{dN} = \left[\left(\frac{-2.115a + 35.72}{100}\right)C_{1100} + \left(1 - \left(\frac{-2.115a + 35.72}{100}\right)\right)C_{2618}\right](\Delta K)^{m_T} \quad (7)$$

The ΔK is the stress intensity factor range defined as [8]:

$$\Delta K = Y \Delta \sigma \sqrt{\pi a} \quad (8)$$

where, Y is a dimensionless function which depends on the geometry and loading configuration and $\Delta\sigma$ is the stress range applied. The Y configuration assumed was half circular surface crack in a shaft [8], and $\Delta\sigma$ was considered constant and was calculated as:

$$\Delta\sigma = \Delta\varepsilon * E \quad (9)$$

where $\Delta\varepsilon$ is the applied strain range and E is the Young's modulus, equal to 74 GPa.

As mentioned before, the multiple linear approximations uses the punctual linear equations (Figure 4) obtained from each test in the calculation of C_T and m_T, and follows the same procedure that the average linear approximation to calculate the total fatigue life. However, the step change occurred differently in each test, for example in the case shown in Figure 4, a will be equal to 5.00 mm.

Ductile fracture contribution found in the fatigue tests carried out at 230°C were include in a similar approach than at room temperature. The fatigue tests were performed at different R due to limitations with the equipment used during this research. C and m values were estimated for each case. A polynomial tendency was used to involve the contribution of the different cracks. The maximum crack length (a) represented by this curve was 5 mm, because the final fracture occurred in the middle of the specimen where the various cracks met. The equation obtained was as follows,

$$\frac{da}{dN}=\left[\left(\frac{-0.961a^2+4.953a+10.27}{100}\right)C_{1100}+\left(1-\frac{(-0.961a^2+4.953a+10.27)}{100}\right)C_{2618}\right](\Delta K)^m \quad (8)$$

MATLAB 7.0 scripts were developed to introduce these models and calculate the fatigue life using a Simpson's rule for numerical integration. An initial crack length (a_o) of 300 μm (approx.10 times the average grain size found in the material) was introduced.

3. Results.

3.1 Room temperature model.

a) Average linear model

In this section the results of the average linear model (equations 6 and 7) will be described. As mentioned before, the best fitting was applying an initial defect (a_o) of 300 μm. The final crack length (a_f) used was 5 mm, because at this point in the test the fatigue life will be consumed.

Figure 5 shows the results found experimentally (diamonds), and were compared with the original equation of Bathias [11] without the DI_F contribution (solid line), and the model developed during this work (dashed line and square mark). Good fitting was found the range between 10^4-10^5 cycles; however at lower fatigue life the model underestimates the number of cycles, while at lower strain amplitudes ($\Delta\varepsilon/2$) the values were overestimated. Table 3 summarized these results. It was noticeable that the DI_F contribution has a significant impact in the final fatigue life.

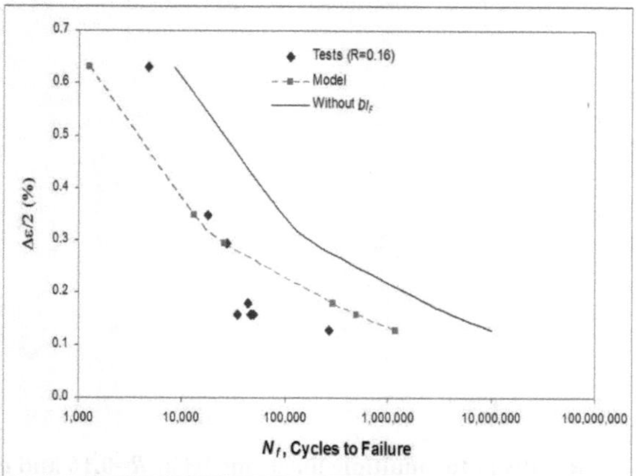

Figure 5. Results of the average linear model at R=0.16 and a_o= 300μm.

Table 3. Summary and comparison of the experimental, without the DI_F contribution and model results.

$\Delta\varepsilon/2$ (%)	$\Delta\varepsilon * E$	N_f Tests (R=0.16)	N_f without DI_F	N_f Model
0.63	932.40	4,666	8,411	1,257
0.35	515.93	18,146	96,327	13,165
0.29	435.12	27,405	194,340	25,886
0.18	266.15	44,170		
0.18	266.15	49,600	2,410,100	292,720
0.16	233.54	34,390		
0.16	233.54	46,500	4,129,500	491,750
0.13	187.66	264,560	10,169,000	1,171,500

b) Multiple linear model

The results using each single DI_F curve for a $\Delta\varepsilon/2$ specified, as shown in Figure 6, are defined as a multiple linear model. The a_o and a_f values were kept the same as in the average linear model. Figure 6 showed the results of this model. The maximum strain amplitude ($\Delta\varepsilon/2$= 0.63%) prediction was improved, which enhanced the shape of the curve. However, an increase in the number of cycles to failure took place in both versions at lower strain amplitude ranges ($\Delta\varepsilon/2$= 0.16% and 0.13%), if these are compared to the single linear model results (Table 3). The results are also expressed in Table 4. The difference between the fatigue life without DI_F contribution and both approximations (linear or multiple) was still considerable, even with the discrepancies in the lower strain amplitudes ranges.

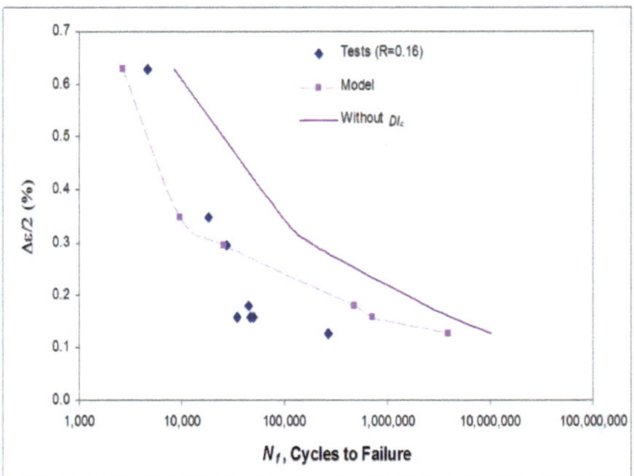

Figure 6. **Results of the multiple linear model at R=0.16 and a_o= 300μm.**

3.2 The 230°C model.

The C and m values used by this approximation were kept the same as room temperature because this data was not available at elevated temperature; however, both C values were changed depending on R. The best a_o found to fit the results was 360 μm, and a_f was consider being 5 mm. The numerical integration of this equation was also made using the Simpson's rule. Figure 7 shows the results achieved using the polynomial approximation using the overall result. As can be seen, the model predicts very well the test points, and the importance of the DI_F contribution is once again recognized. The fatigue life is almost 10 times reduced. Table 5 compares the overall results.

Table 4. Summary of the multiple linear model results.

$\Delta\varepsilon/2$ (%)	$\Delta\varepsilon * E$	N_f Tests (R=0.16)	N_f without DI_F	N_f Model
0.63	932.40	4,666	8,411	2,677
0.35	515.93	18,146	96,327	9,618
0.29	435.12	27,405	194,340	25,482
0.18	266.15	44,170	2,410,100	478,120
0.18	266.15	49,600		
0.16	233.54	34,390	4,129,500	715,390
0.16	233.54	46,500		
0.13	187.66	264,560	10,169,000	3,951,900

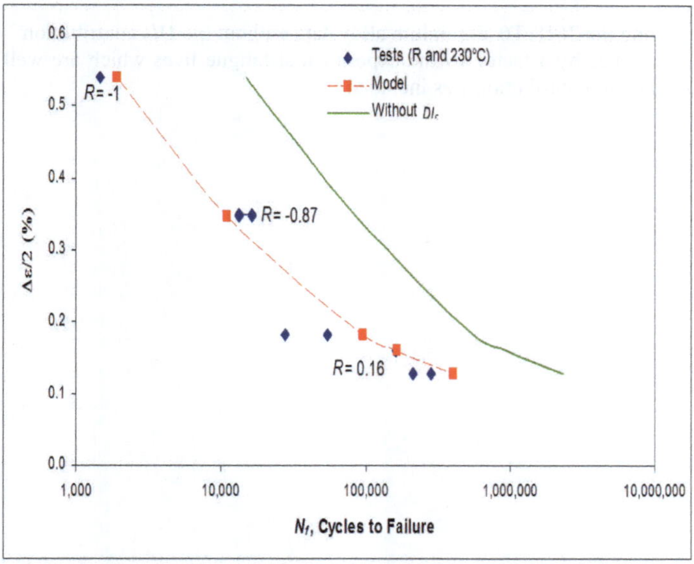

Figure 7. Results of the 230°C model using the average polynomial equation.

Table 5. **Results using single polynomial model and respectively *R*.**

$\Delta\varepsilon/2$ (%)	$\Delta\varepsilon * E$	*R*	N_f Tests	N_f Without DI_F	N_f Model
0.54	793.80	-1	1,472	15,104	2,004
0.35	512.44	-0.87	13,492 - 16,602	85,652	11,029
0.18	267.96	0.16	54,685	556,740	87,190
0.18	267.96	0.16	27,930		
0.16	235.20	0.16	162,090	952,750	146,830
0.13	189.00	0.16	215,370	2,345,700	351,880
0.13	189.00	0.16	284,740		

4. Conclusions

The main conclusions of this work are listed below:

1. The two-phase material model developed during this research which included the DI_F contribution in the Paris equation parameters, *C* and *m*, shows the importance of taking into account the modes of fracture in the fatigue life prediction.

2. The multiple linear approximations which used the DI_F curve of each test gives the best fatigue life prediction at room temperature.

3. Modelling the 2618-T6 aluminium alloy data without the DI_F contribution overestimates by a factor 10 the experimental fatigue lives which are well predicted when the DI_F contribution was included.

5. References

[1] PARDOEN, T., DUMONT, D., DESCHAMPS, A. & BRECHET, Y. (2003) Grain boundary versus transgranular ductile failure. Journal of the Mechanics and Physics of Solids, 51, 637-665.

[2] DE HAAS, M. & DE HOSSON, J. T. M. (2002) On the effects of thermomechanical processing on failure mode in precipitation-hardened aluminium alloys. Journal of Materials Science, 37, 5065-5073.

[3] GANDIN, C. A. & JACOT, A. (2007) Modeling of precipitate-free zone formed upon homogenization in a multi-component alloy. Acta Materialia, 55, 2539-2553.

[4] DUMONT, D., DESCHAMPS, A., BRÉCHET, Y., SIGLI, C. & EHRSTRÖM, J. C. (2004) Characterisation of precipitation microstructures in aluminium alloys 7040 and 7050 and their relationship to mechanical behaviour. Materials Science and Technology, 20, 567-576.

[5] JIANG, H. & FAULKNER, R. G. (1996) Modelling of grain boundary segregation, precipitation and precipitate-free zones of high strength aluminium alloys-I. The model. Acta Materialia, 44, 1857-1864.

[6] JIANG, H. & FAULKNER, R. G. (1996) Modelling of grain boundary segregation, precipitation and precipitate-free zones of high strength aluminium alloys-II. Application of the models. Acta Materialia, 44, 1865-1871.

[7] SALAS ZAMARRIPA, A., PINNA, C., BROWN, M.W., GUERRERO MATA, M.P., CASTILLO MORALES, M. & BEBER-SOLANO, T.P. (2011) Identification of modes of fracture in a 2618-T6 aluminum alloy using stereophotogrammetry. Materials Characterization, 62, 1141-1150

[8] DOWLING, N. E. (1993) Mechanical Behaviour of Materials, Prentice-Hall International Editions.

[9] SURESH, S. (1998) Fatigue of Materials, CAMBRIDGE University Press.

[10] PARIS, C. P. (1998) Fracture mechanics and fatigue: a historical perspective. Fatigue & Fracture of Engineering Materials & Structures, 21, 535-540.

[11] BATHIAS, C. (1978) Application of fracture mechanics to aluminum alloys selection. Engineering Fracture Mechanics, 10, 267-282.

[12] BILIR, O. G. & HARUN, M. (1990) Effect of stress ratio on the rate of growth of fatigue cracks in 1100 Al-alloy. Engineering Fracture Mechanics, 37, 1203-1206.

[13] MATWEB (2004), http://www.matweb.com/. 2618-T6 Material Property Data. Accessed: March 2004

[14] WANGA, J., YIB, D, SUA, X. & YINA, F. (2007) Influence of deformation ageing treatment on microstructure and properties of aluminum alloy 2618. Materials Characterization, 59, 965–968.

6. Acknowledgments

The authors gratefully acknowledge the financial support of CONACyT-México (Consejo Nacional de Ciencia y Tecnología) and to Cummins Turbo Technologies for providing the raw material and their technical support. Special thanks to Prof. Qiang Zhu for his supportiveness during the entire project.

Fatigue of Materials III: Advances and Emergences in Understanding
Edited by: T.S. Srivatsan, M. Ashraf Imam, and R. Srinivasan
TMS (The Minerals, Metals & Materials Society), 2014

FATIGUE CRACK GROWTH IN PRE-STRESSING STEEL WIRES: TRANSIENT AND STEADY-STATE REGIMES

J. Toribio[1], J.C. Matos[2], B. González[1]

[1] Department of Materials Engineering
University of Salamanca
E.P.S. Zamora, **Spain**

[2] Department of Computing Engineering
University of Salamanca
E.P.S. Zamora, **Spain**

Abstract

This paper analyzes the propagation of fatigue cracks in pearlitic steel supplied in two forms, hot rolled bar and prestressing steel wire. The experimental procedure consisted of fatigue tests on cylindrical bars under tensile loading, using steps with decreasing amplitude of stress intensity factor and constant stress intensity range during each step. The curves plotting crack growth rate versus stress intensity rate show a main steady-state regime preceded by different transient paths. The steady-state regime is associated with the curves of the Paris regime. For a given material, the crack growth rate is not dependent on the *R*-ratio. With regard to the comparison between the fatigue behaviour of the hot rolled bar and the prestressing steel wire, a retardation is observed in the latter. The transient branches allow one to calculate the plastic zone size, considering that they are a consequence of the overload retardation effect at each step change.

Keywords: Cold-drawn steel, Paris curve, R-ratio, Plastic zone size, Overload retardation effect

1. Introduction

Fatigue phenomenon is usually considered as a two load parameter problem [1-3] involving two crack tip driving forces, the stress intensity factor range and the maximum stress intensity factor (ΔK and Kmax).

In fatigue crack growth of fully pearlitic steels, the increase of the load ratio, R-ratio, produces a significant raise of the slope of the Paris-Erdogan curve; this phenomenon coincides with an increase in the amount of cleavage fracture present in the fracture surface [4]. The drawing process of eutectoid steel induces retardation in the propagation of fatigue cracks, due to the microstructural changes which produce a rougher microcracking [5].

The main reasons of the retardation in fatigue advance caused by overload are crack closure due to the residual compressive stresses, crack branching and contact between the rough crack surfaces after overload [6]. The retarding effect due to consecutive overloads is found to increase with the number of overloads, until it reaches a maximum [7].

There are several models to calculate the plastic zone size of the crack tip: by means of a simple application of the linear elastic fracture mechanics [8], with the strip yield model [9,10], using the superposition method [11] or according to the condition of positive equivalent plastic strain rate [12,13].

The aim of the present paper is to study the growth of fatigue cracks in a high strength steel supplied as a hot rolled bar (base material) and as a cold drawn wire (commercial product), by means of fatigue tests using steps with decreasing stress amplitude, through the analysis of both the transient and steady-state regimes in the cyclic crack growth curve da/dN *vs.* ΔK.

2. Experimental Procedures

The material was pearlitic steel, whose composition appears on Table I, supplied in two forms: firstly, as a hot rolled bar which has not been cold drawn at all and, secondly, as a commercial cold drawn wire which has undergone seven steps of cold drawing and a thermo- mechanical treatment to eliminate (or at least relieve) residual stresses at its surface.

Table I. **Chemical composition (wt %) of the steels.**

% C	% Mn	% Si	% P	% Cr	% V
0.789	0.681	0.210	0.010	0.218	0.061

The mechanical properties of both steels were obtained after a standard tension test (Figure 1). Table II shows the Young's modulus E, a factor that barely changes with the drawing process, the 0.2 % conventional yield strength σY and the tensile strength σR, the two latter increasing notably after drawing.

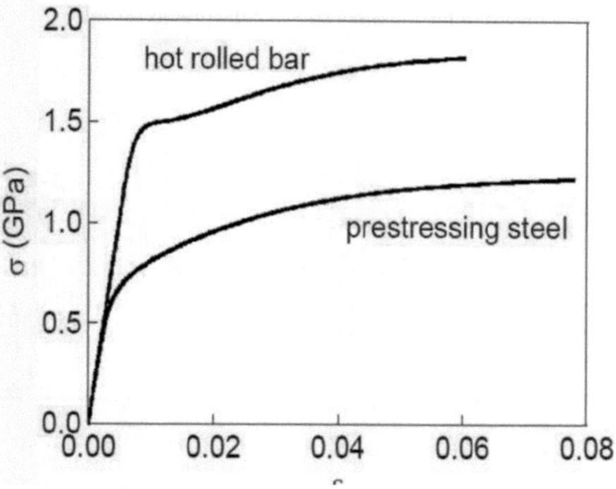

Figure 1. Curve σ-ε (standard tension tests) for the hot rolled bar and the cold drawn wire.

Table II. Mechanical properties of the steels.

Steel	E (GPa)	σ_Y (GPa)	σ_R (GPa)	ε_R
Hot rolled bar	202	0.70	1.22	0.08
Cold drawn wire	209	1.48	1.82	0.06

Transversal and longitudinal samples from each material form were cut, mounted, grounded and polished until a mirror surface was obtained. After being etched with 4 % Nital (mixture of 4 ml of nitric acid with 96 ml of ethanol) to reveal microstructure, they were examined by scanning electron microscopy (SEM) with magnification factor up to x6000.

In fatigue tests the specimens used were cylindrical bars of 300 mm of length, diameters of 11.0 mm and 5.1 mm for the hot rolled bar and for the cold drawn wire, respectively. Before the test, a small transversal cut was performed on the middle part of the specimens, so as to cause the crack initiation at this point. During the test, the applied load F, the extensometer displacement u (placed on the specimen in a symmetric way to the crack front) and the number of cycles N, were recorded.

The test procedure consisted of applying an oscillating tensile loading in the axial direction in successive decreasing steps, with a constant stress range $\Delta\sigma$ during each step and decreasing value from one to another step. The frequency used was 10 Hz, with the shaped of a sinusoidal wave and different R-ratios: 0, 0.25 and 0.50 (Figure 2). The initial maximal stress was always lower than the yield strength, decreasing in the next steps around 20÷30 % with regard to the maximal stress in the previous step.

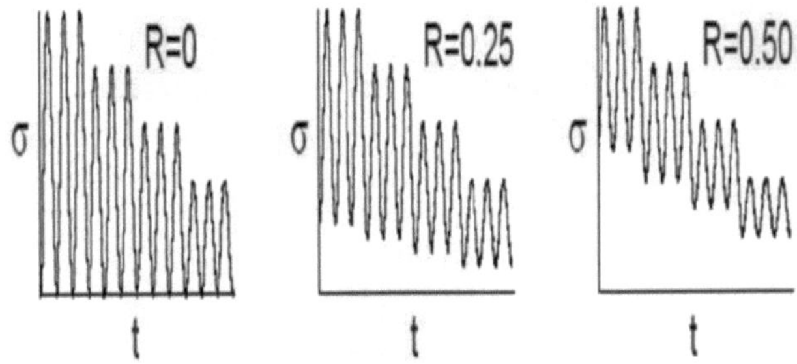

Figure 2. Loading cycles in the different fatigue tests.

Every load step was maintained enough time to be able to appreciate the crack advance and eliminate the plastic effect at the crack tip caused by the previous step. The tests allowed one to characterize the evolution of the crack front for both steels (hot rolled bar and cold drawn wire) and obtaining the fatigue crack growth curve for the different *R*-ratios.

3. Experimental Results

After the tests, the fatigue surfaces of the specimens were observed (Figure 3). In both steels the fatigue fracture surface was contained in the transverse section of the wire (perpendicular to the loading direction), thereby exhibiting mode I fracture even in the prestressing steel wire (heavily drawn) where a microstructural anisotropy does exist due to the cold drawing process.

Figure 3. Fatigue fracture surface, hot rolled bar (left) and prestressing steel wire (right).

In every load change the crack front was modelled as part of an ellipse with centre on the bar surface, of semiaxes a (crack depth) and b, from a set of points taken on that front and using the least squares method, following a procedure described by Toribio *et al.* [14].

In the central point of the crack front, where a plane strain state is achieved due to the enough constraint there, the curve representing the fatigue crack growth was calculated: cyclic crack growth rate da/dN versus stress intensity factor range ΔK. The value of the ΔK for the studied geometry, the crack size and the type of applied load, is:

$$\Delta K = Y \Delta\sigma \sqrt{\pi a} \qquad (1)$$

The dimensionless stress intensity factor (SIF) Y, calculated by Astiz [15], for the central point of the crack front by using the finite element method, was used depending on the relative crack depth (crack depth divided by the diameter, a/D) and the aspect ratio (ratio between the semiaxes of the ellipse, a/b),

$$Y = \sum_{\substack{i=0 \\ i\neq 1}}^{4} \sum_{j=0}^{3} C_{ij} \left(\frac{a}{D}\right)^{i} \left(\frac{a}{b}\right)^{j} \qquad (2)$$

with the Cij coefficients on Table III.

Table III. Coefficients Cij of Equation (2).

i	*j=0*	*j=1*	*j=2*	*j=3*
0	1.118	-0.171	-0.339	0.130
2	1.405	5.902	-9.057	3.032
3	3.891	-20.370	23.217	-7.555
4	8.328	21.895	-36.992	12.676

The procedure followed to obtain the fatigue crack growth curves was that outlined in Figure 4. First of all, the a/D *vs.* $1/CED$ plot (where C is flexibility, $C=\Delta u/\Delta F$) and the a/b *vs.* a/D plot were calculated from the characterization of the crack front, obtained when the load was changed, and the F-u data, obtained on the test registry, and both curves were fitted providing polynomial equations of the third degree with high regression coefficients. These fit curves were used to calculate in every instant, from the value of the dimensionless stiffness $1/CED$, the relative crack depth a/D and from this, the value of the aspect ratio a/b, so that the geometry characterization of the crack front is available at every single moment of the test.

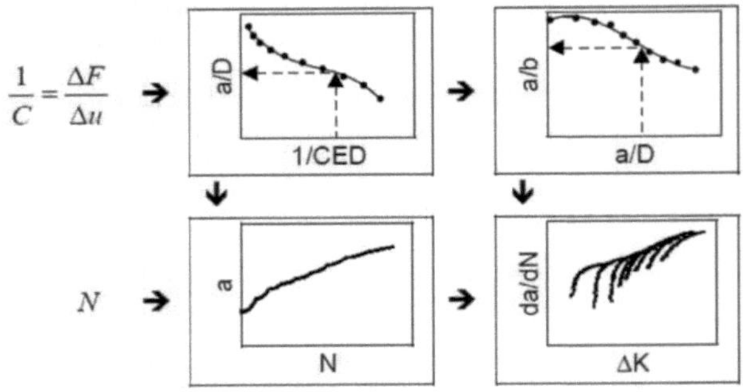

Figure 4. Calculation of the fatigue crack growth rate.

Finally, the cyclic crack growth rate da/dN (deducted from the a-N curve) *versus* the SIF range ΔK (calculated using the expression provided by Astiz) was obtained, where every load step constitutes a different section of the curve, converging all of them asymptotically to a common Paris branch. These curves for the hot rolled bar and the cold drawn wire are shown in Figure 5.

In every step, where $\Delta\sigma$ is constant, there is an increase on the value of the SIF range. On the contrary, the load's sudden decrease when changing from one step to the following implies a reduction in the value of the SIF range, as every stage starts with a smaller SIF than the last value of the previous phase.

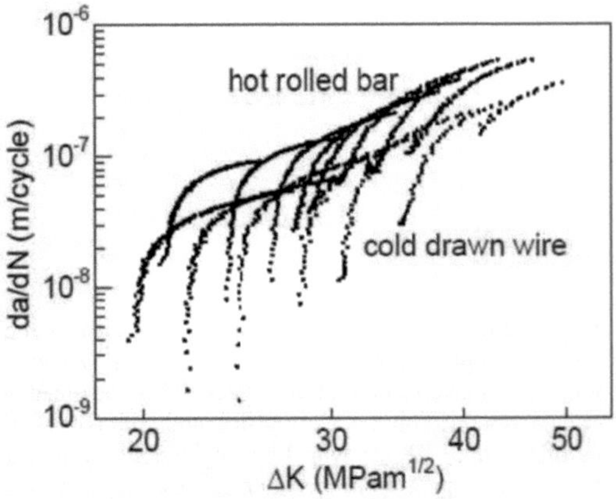

Figure 5. da/dN-ΔK curve: transient paths and steady-state regime.

4. Discussion

The sudden decrease of the load when changing from one to another step is probably the cause of the appearance of a transient path in the da/dN-ΔK curve, where the rate shows a sort of retardation on the advance (which is the cause of the branches on the graphic representation in Figures 4 and 5). This decrease is caused by the overstress which appears when changing from a greater load increase to a lesser one, according to the mechanism of overload retardation generated by the plastic zone close to the crack tip [16,17]. Once this zone has been crossed, the rate corresponding to the steady-state regime is reestablished.

4.1 Steady-State Regime vs. Paris Law

Figure 6 plots the da/dN-ΔK curve in the steady-state region (Paris regime) for both the hot rolled bar and the cold drawn wire, with R-ratios of 0, 0.25 and 0.50.

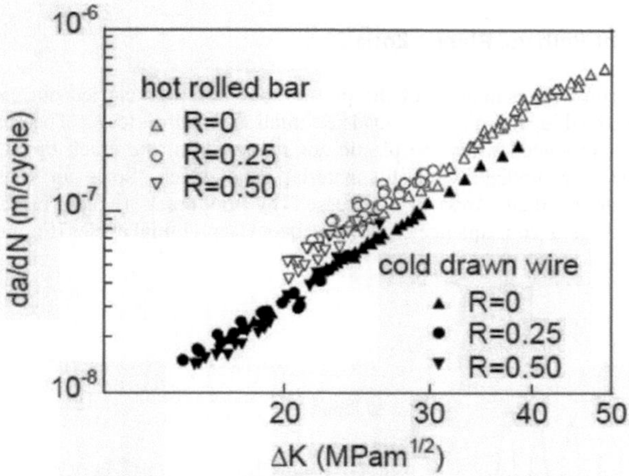

Figure 6. Paris regime for the hot rolled bar and the cold drawn wire.

For each steel and for different values of the R-ratio, there is a unique fit of the fatigue crack growth rate curve (da/dN *vs.* ΔK) in the Paris regime, i.e., for a given material there is a single Paris law plot with independence of the R-ratio (it implies a unique C and m coefficients for each steel under distinct R-ratios),

$$\frac{da}{dN} = C\Delta K^{m} \qquad (3)$$

Although there is shifting of the plot towards lower values of ΔK when the R-ratio increases. Furthermore, the bilogarithmic fit is parallel for both steels, hot rolled bar and cold drawn wire, with the same slope and, therefore, the same Paris exponent m, but the cold drawn wire curve appears down and to the left (with a lower Paris coefficient C) in relation to that of the hot rolled bar (Table IV). Therefore, the cold drawn prestressing steel wire exhibits fatigue crack growth retardation in relation to the behaviour of the hot rolled bar, i.e., there is an improvement of fatigue performance in the former.

Table IV. Paris coefficients C and m (units for da/dN in m/cycle and ΔK in MPa $m^{1/2}$).

Steel	m	C
Hot rolled bar	3.0	$5.3 \cdot 10^{-12}$
Cold drawn wire	3.0	$4.1 \cdot 10^{-12}$

4.2 Transient Path vs. Plastic Zone Size

An experimental estimation of the plastic zone size was carried out, assuming that this is the responsible for the overload retardation in the tests. The retardation is considered to be produced when the plastic zone emerged at the crack tip due to the last cycles of a step (surrounded by elastic material zone which "sews up" the crack with high compressive residual stresses) is crossed by the crack (which presents a small plastic zone size) as a consequence of the subsequent step's initial cycles (Figure 7).

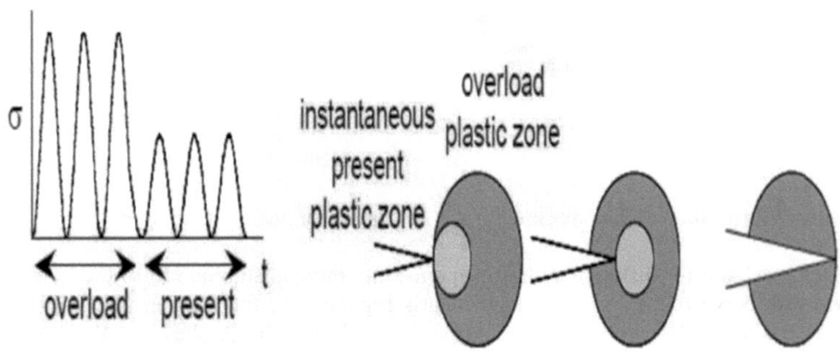

Figure 7. Propagation through the plastic zone due to overload.

The overload retardation effect has been proved to depend on the number of cycles which cause it [7], although it is considered that the last load cycle is mainly responsible.

An expression to calculate the plastic zone size, which is similar to that proposed by other authors [8-10], has been studied. Taking into account that the plastic zone size depends on the Kmax value but also of the R-ratio, a dependency on the KmaxΔK product as the best solution is proposed, since it considers both parameters governing the mechanisms in the vicinity of the crack tip. Fatigue phenomenon is usually considered as a two load parameter problem [1-3] involving two crack tip driving forces, the stress intensity factor range and the maximum stress intensity factor (ΔK and Kmax).

Furthermore, in pearlitic steel, material with strain hardening, the yield strength σY and the tensile strength σR have been taken into account in the plastic zone size expression's denominator, both having been obtained on the standard tension test (Figure 1). The proposed expression to calculate the plastic zone size rp was (where k is a fitting coefficient):

$$r_p = k\frac{K_{max}\Delta K}{\left(\frac{\sigma_Y+\sigma_R}{2}\right)^2} \quad (4)$$

In Figure 8 the experimental results of the plastic zone size, for the hot rolled bar and the cold drawn wire, with different values of the R-ratio (0, 0.25 and 0.50), are shown. It is observed that the plastic zone size is greater in the hot rolled bar (with smaller yield strength and tensile strength) than in cold drawn wire, deducting for both materials a unique fit for the plastic zone size, which presents a k coefficient of 0.14, similar to that obtained by other authors [8-10].

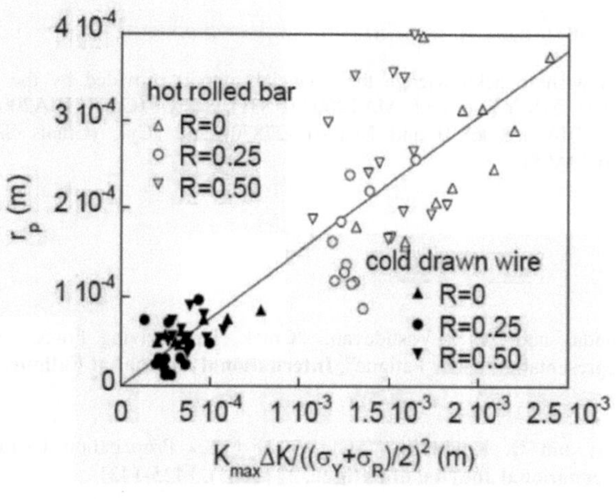

Figure 8. ***Plastic zone size for the hot rolled bar and the cold drawn wire.***

5. Conclusions

(i) The cyclic crack growth curves obtained from the fatigue tests in successive steps, with constant $\Delta\sigma$ in every step and decreasing regarding the previous one, are formed by sections which converge asymptotically to a common Paris branch, showing steady-state regime and transient paths.

(ii) The change of the R-ratio (with positive values between 0 and 0.5) does not alter the cyclic crack growth rate, because the Paris curve is the same (the da/dN-ΔK plotting for the different R-ratios have the same fit equation), although its values tend to lower ΔK

when R-ratio increases.

(iii) The cold drawing process improves the fatigue behavior of eutectoid steel by retarding the cyclic crack growth rate on the Paris regime (the Paris line of the cold drawn wire moves down and to the left regarding that of the hot rolled bar, maintaining itself parallel).

(iv) In the estimation of the plastic zone size on the basis of overload retardation, apart from the conventional mechanical properties of the material (σY and σR), two fracture mechanics parameters (ΔK and Kmax representing the two key driving forces) have to be considered to completely characterize the fatigue process.

(v) The plastic zone size is higher for the hot rolled bar than for the cold drawn prestressing steel wire where the drawing process raises the yield stress and the ultimate tensile strength. The proposed relationship to calculate the plastic zone size, similar to that obtain by other authors, has a fitting coeffcient of about 0.14.

Acknowledgements

The authors wish to acknowledge the financial support provided by the following Spanish Institutions: MICYT (Grant MAT2002-01831), MEC (Grant BIA2005-08965), MICINN (Grants BIA2008-06810 and BIA2011-27870) and JCyL (Grants SA067A05, SA111A07 and SA039A08).

References

1. K. Sadananda, and A.K. Vasudevan, "Crack Tip Driving Forces and Crack Growth Representation under Fatigue", **International Journal of Fatigue**, 26 (2004), 39-47.

2. S. Stoychev, and D. Kujawski, "Analysis of Crack Propagation Using ΔK and Kmax", **International Journal of Fatigue,** 27 (2005), 1425-1431.

3. J. Zhang, X.D. He, and S.Y. Du, "Analyses of the Fatigue Crack Propagation Process and Stress Ratio Effects Using the Two Parameter Method", **International Journal of Fatigue**, 27 (2005), 1314-1318.

4. A.B. El-Shabasy, and J.J. Lewandowski, "Effects of Load Ratio, R, and Test Temperature on Fatigue Crack Growth of Fully Pearlitic Eutectoid Steel (Fatigue Crack Growth of Pearlitic Steel)", **International Journal of Fatigue**, 26 (2004), 305-309.

5. J. Toribio, J.C. Matos, and B. González, "Micro- and Macro-Approach to the Fatigue Crack Growth in Progressively Drawn Pearlitic Steels at Different R-ratios", **International Journal of Fatigue**, 31 (2009), 2014-2021.

6. S. Suresh, "Micromechanisms of Fatigue Crack Growth Retardation Following Overloads", **Engineering Fracture Mechanics**, 18 (1983), 577-593.

7. S. Pommier, and M. de Freitas, "Effect of Fatigue Crack Growth of Interactions Between Overloads", **Fatigue & Fracture of Engineering Materials & Structures**, 25 (2002), 709-722.

8. G.R. Irwin, "Plastic Zone Near a Crack and Fracture Toughness", Mechanical and Metallurgical Behaviour of Sheet Materials (New York: Syracuse University, 1960), 63-71.

9. D.S. Dugdale, "Yielding of Steel Sheets Containing Slits", **Journal of the Mechanics of Physics and Solids**, 8 (1960), 100-104.

10. G.I. Barenblatt, "The Mathematical Theory of Equilibrium Cracks in Brittle Fracture",**Advances in Applied Mechanics**, 7 (1962), 55-129.

11. J.R. Rice, "Mechanics of Crack Tip Deformation and Extension by Fatigue", Fatigue Crack Propagation —ASTM STP 415 (West Conshohocken: ASTM International, 1967), 247-311.

12. J. Toribio, and V. Kharin, "Role of Crack Tip Mechanics in Stress Corrosion Cracking of High-Strength Steels", **International Journal of Fracture**, 126 (2004), L57-L63.

13. J. Toribio, and V. Kharin, "Fractographic and Numerical Study of Hydrogen-Plasticity Interactions Near a Crack Tip", **Journal of Materials Science**, 41 (2006), 6015-6025.

14. J. Toribio, J.C. Matos, B. González, and J. Escuadra, "An Automated Procedure for the Geometrical Modelling of a Surface Crack Front", Structural Durability and Health Monitoring, 123 (2009), 1-16.

15. M.A. Astiz, "An Incompatible Singular Elastic Element for Two- and Three-Dimensional Crack Problems", **International Journal of Fracture**, 31 (1986), 105-124.

16. S. Suresh, Fatigue of Materials (Cambridge: Cambridge University Press, 1994).

17. S.Y. Yarema, Test Method for the Determination of Crack Growth Rates and Crack Growth Resistance under Cyclic Loading (L'viv: Ukrainian Academy of Sciences, 1994).

Fatigue of Materials III: Advances and Emergences in Understanding
Edited by: T.S. Srivatsan, M. Ashraf Imam, and R. Srinivasan
TMS (The Minerals, Metals & Materials Society), 2014

EVOLUTION OF CRACK ASPECT RATIO IN SHEETS UNDER TENSION AND BENDING CYCLIC LOADING

J. Toribio[1], J.C. Matos[2], B. González[1], J. Escuadra[2]

[1] Department of Materials Engineering,
University of Salamanca,
E.P.S. Zamora, Spain

[2] Department of Computing Engineering,
University of Salamanca,
E.P.S. Zamora, Spain

Abstract

The aim of this paper is the numerical prediction of the cracking path followed by a surface crack front in plates constituted of different materials (determined by the exponent m of the Paris law), subjected to cyclic tension or cyclic bending loading. To this end, a numerical modelling was developed on the basis of a discretization of the crack front (characterized with elliptical shape) and the crack advance at each point perpendicular to such a front, according to a Paris law, using the stress intensity factor (SIF) calculated by Newman and Raju. Results show that the crack leads to a preferential crack path that corresponds to a very shallow initial crack with a quasi-circular crack front. The increase of the Paris exponent produces a quicker convergence during fatigue crack propagation from the different initial crack shapes.

Keywords: Numerical modelling, Cracked plate, Paris law, Crack advance, Crack aspect ratio

1. Introduction

The plate, a simple geometry of structural application, often presents cracks on its surface that can easily propagate due to fatigue, a reason why studying them is of great interest for Fracture Mechanics. The stress intensity factor (SIF) in a superficially cracked plate, subjected to tension or bending loads, has been calculated by several methods [1-8]. For tension loading the SIF increases with crack depth, showing its maximum value at the centre of the front for semi-elliptical cracks (with a ratio between semiaxes 0.2) and at the point of the surface for semicircular cracks [1,2]. For bending moment the SIF is highest at the crack surface from relative depths greater than or equal to 0.6, regardless of the geometry of the crack front [1].

The advance of the crack front in plates has been experimentally [1,8-13] and numerically [1,6-8,10-13] analyzed, applying the Paris Erdogan law [14]. Cracks try to grow so that the SIF remains constant along the front (iso-*K* condition of propagation), but the plate's free surfaces and the bending load prevent this [15]. Cracks asymptotically tend to a preferential propagation path [1,6-8,10-13], so that the larger the ratio of the bending loading to tension loading, the more intense the asymptotic tendency [12]. If the Paris exponent is greater, the change in the aspect ratio is more pronounced [12].

2. Numerical Modelling

A computer program in Java programming language was developed to study the propagation path of a symmetric surface crack on the cross section of a plate (Figure 1), subjected to fatigue (cyclic) loading in the form of remote uniform tension or pure bending moment (under Mode-I loading conditions).

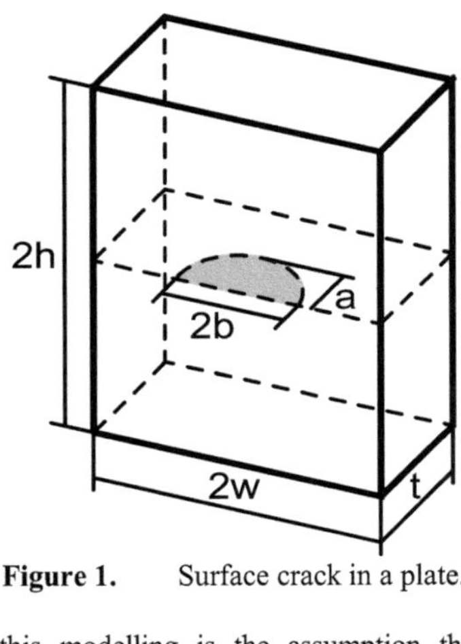

Figure 1. Surface crack in a plate.

The basic hypothesis of this modelling is the assumption that the crack front can be characterized as a semi-ellipse with its center on the plate's surface and that fatigue propagation takes place in a direction perpendicular to such a front, following the Paris Erdogan law [14],

$$\frac{da}{dN} = C\Delta K^{m} \tag{1}$$

To discretize the crack front, modelled in a semi-elliptical way, the front was divided into z segments of equal length, using Simpson's rule. The point on the surface of the plate, where there is plane stress, was not taken into account. Subsequently, each of the points (i) was moved perpendicularly to the crack front, in accordance with the Paris Erdogan law, so that the maximum crack increase, $\Delta a(\max)$, corresponding to the point of the maximum SIF, $Y(\max)$, was kept constant. From this maximum increase and dimensionless SIF Y (where K is the SIF and σ remote tension stress or maximum remote bending stress),

$$Y = \frac{K}{\sigma\sqrt{\pi a}} \tag{2}$$

the advance of each of the front points can be obtained Δa_i,

$$\Delta a_i = \Delta a(\max)\left[\frac{Y_i}{Y(\max)}\right]^{m} \tag{3}$$

The new points, fitted by the method of least squares, form a new semi-ellipse, so that the crack advancement progress is repeated iteratively. Due to the existing symmetry, only half of the problem was used in the calculations. The parameters z and $\Delta a(\max)$ were determined by a study of convergence.

The SIFs used were those obtained by Newman and Raju [1,2] by a 3D finite element analysis and the nodal-force method. These authors adjusted their results to an equation [1,13] that calculates the dimensionless SIF with parameters a/t, a/b, and ϕ, which characterizes each point of the crack front (Figure 2). This equation (valid for $0 < a/b \leq 1.0$, $0 \leq a/t < 1.0$, $0 \leq \phi \leq \pi$ and $b/w < 0.5$) has been tested experimentally and numerically by various methods [1,4,7-11,13,15]: measurements of K (photoelastic and numerical), the geometrical evolution of the front, the crack growth rate, failure stress, etc. Although other equations do exist, this one has been the most widely used in the bibliography due to its excellent results.

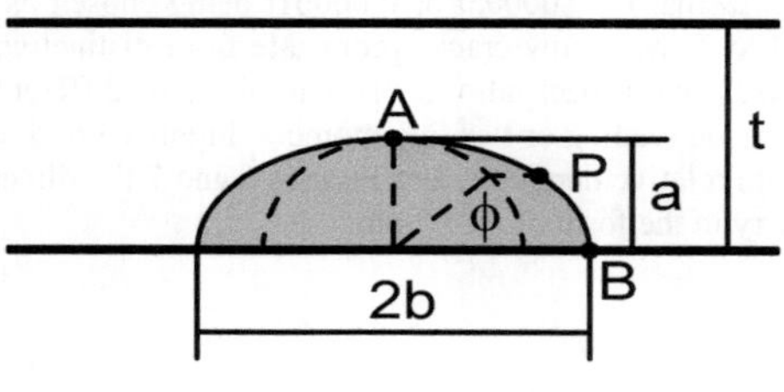

Figure 2. Parameters that characterize a point P at the crack front (a/t, a/b, ϕ).

Keeping in mind Newman and Raju's equation [1,13], the advancement at each point of the crack front can be calculated using the expression,

$$\Delta a_i = \Delta a(\max)\left[\frac{(Hf_\phi g)_i}{(Hf_\phi g)_{\max}}\right]^m \tag{4}$$

wherein the parameters f_ϕ, g, and H are obtained by the following equations,

$$f_\phi = \left[\left(\frac{a}{b}\right)^2 \cos^2\phi + \sin^2\phi\right]^{1/4} \tag{5}$$

$$g = 1 + \left[0.1 + 0.35\left(\frac{a}{t}\right)^2\right](1 - \sin\phi)^2 \tag{6}$$

The parameter H is equal to 1 in the case of tension loading, while for the bending loading:

$$H = H_1 + (H_2 - H_1)\sin^p\phi \tag{7}$$

$$p = 0.2 + \frac{a}{b} + 0.6\frac{a}{t} \tag{8}$$

$$H_1 = 1 - 0.34\left(\frac{a}{t}\right) - 0.11\left(\frac{a}{b}\right)\left(\frac{a}{t}\right) \tag{9}$$

$$H_2 = 1 + \left[-1.22 - 0.12\left(\frac{a}{b}\right)\right]\left(\frac{a}{t}\right) + \left[0.55 - 1.05\left(\frac{a}{b}\right)^{0.75} + 0.47\left(\frac{a}{b}\right)\right]^{1.5}\left(\frac{a}{t}\right)^2 \tag{10}$$

2. Numerical Results and Discussion

A convergence study was carried out in order to determine the number of parts in which each semi-ellipse is divided (crack front) and the maximum crack depth in the iterations, $z = 24$ and $\Delta a(\max) = 0.00005t$ or $0.00001t$ being chosen as was required by the calculation. Figures 3 to 7 show how cracks propagate from distinctive initial geometry — relative crack depth $(a/t)_0$ and aspect ratio $(a/b)_0$— in plates of different materials (m = 2, 3 and 4), subjected to tension loading or bending moment. Figures 3 to 5 show the evolution of the aspect ratio a/b with relative depth a/t, and Figures 6 and 7 the dimensionless advance of the crack front geometry in the form of a/t-b/t plot.

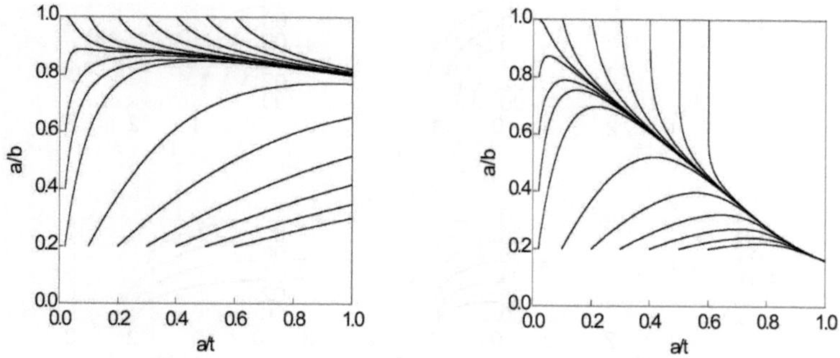

(a) (b)

Figure 3. Crack aspect ratio evolution for *m*=2
(a) tension loading, and (b) bending moment.

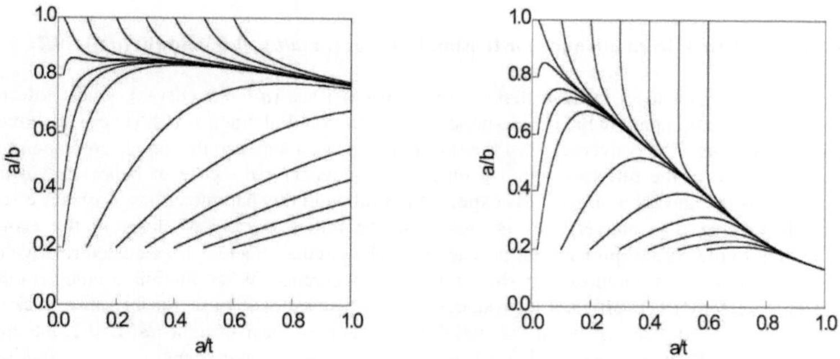

(a) (b)

Figure 4. Crack aspect ratio evolution for *m*=3:
(a) tension loading, and (b) bending moment.

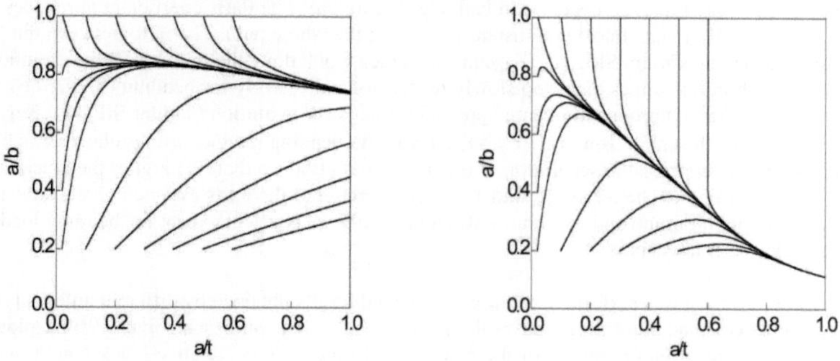

(a) (b)

Figure 5. Crack aspect ratio evolution for m=4:
(a) tension loading, and (b) bending moment.

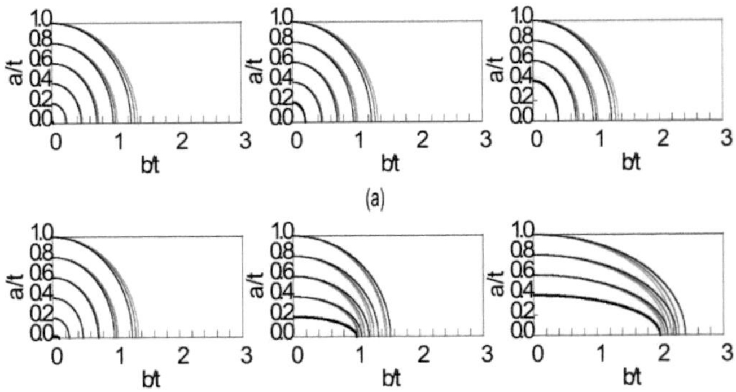

(b)

Figure 6. Crack front advance for tension loading: (a) $(a/b)_0$=1.0, and (b) $(a/b)_0$=0.2.

The crack advance from different initial geometries (*a/b-a/t* curves) tends towards a preferential propagation path, which corresponds to that of an initial crack which is very shallow with a quasi-circular front. The convergence of the solution (approach between the curves corresponding to the propagation from the different initial geometries) is higher in the case of bending than in the situation of tension, and for a larger Paris exponent m (although this parameter has a weaker effect on the growth pattern). The convergence is greater as the initial crack gets closer to the geometry corresponding to the path of preferential propagation. The circular cracks with elevated relative crack depth for tension loading are those that show a lesser convergence. When the initial crack is circular $(a/b)_0$=1.0, a greater Paris coefficient m produces lower aspect ratios (that is, for the same crack depth a, bigger semi-axes b). Conversely, if the initial aspect ratio is quasi-straight $(a/b)_0$~0.2, the curves' behaviour with the Paris exponent m depends on the initial relative crack depth $(a/t)_0$ and the type of stress (tension loading or bending moment).

The preferential propagation path shows a slight decrease in the aspect ratio with the relative crack depth in tension and a marked decrease in bending (Figure 8a). The Paris coefficient m reduces these curves slightly in an almost equidistant way along the whole path. The ratio between the minimum and the maximum SIF, Y_{min}/Y_{max}, in each crack front along the preferential propagation path shows high initial values that drop slowly for tension and sharply for bending (Figure 8b). In tension, the preferential propagation path approaches the iso-K evolution (similar SIF between different points of the crack front, i.e., $Y_{min} \cong Y_{max}$) while in bending the increasing relative crack depth produces an appreciable separation of this iso-K evolution, so that, the higher the crack depth, the greater the difference between Y_{min} and Y_{max}. The increase of the Paris exponent m elevates the ratio between the minimum and maximum SIF, a generally weak effect except for bending loading and high relative crack depths

A comparison was performed of the experimental results obtained by different authors [16,17] and the numerical modelling proposed in the present paper. It is shown in Figure 9 that plots the fatigue crack propagation (evolution of the crack aspect ratio, *a/b*, *vs.* relative crack length, *a/t*) for both tension loading and bending moment. Both plots exhibit an excellent agreement between numerical predictions (solid lines) and experimental results (dotted lines), in spite of the inherent scattering of any fatigue phenomena.

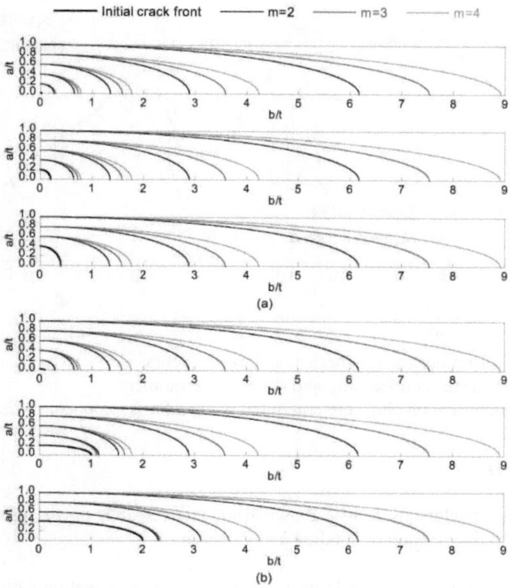

Figure 7. Crack front advance for bending moment: (a) $(a/b)_0$=1.0, and (b) $(a/b)_0$=0.2.

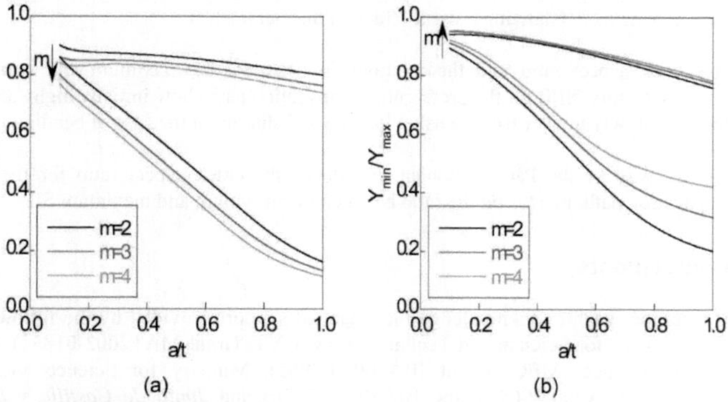

Figure 8. Preferential cracking path: (a) crack aspect ratio, and (b) minimum and maximum SIF ratio.

7

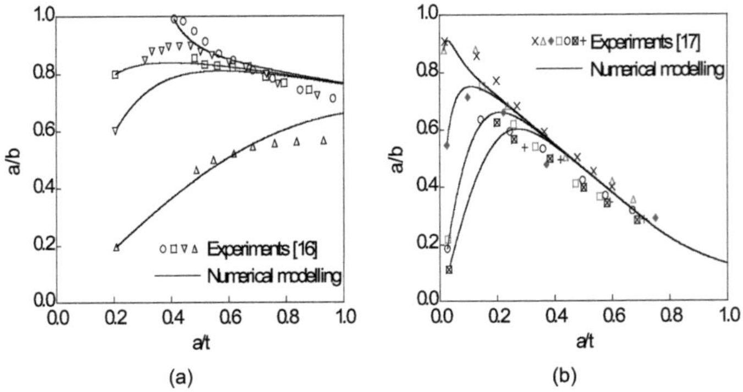

Figure 9. Comparison between numerical and experimental results: (a) tension loading, and (b) bending moment. The numerical results were computed using a Paris exponent m=3.

3. Conclusions

The following conclusions can be drawn from this research on the propagation of surface cracks in plates subjected to tension or bending cyclic loading:

(i) The cracks in fatigue growth tend towards a preferential propagation path, which corresponds to that of an initial crack which is very shallow with a quasi-circular front.

(ii) The convergence, approaching between the propagation curves with different initial geometries, is higher for cyclic bending loading than for cyclic tension loading and for greater Paris exponents *m* (showing a weak effect on this parameter).

(iii) The crack aspect ratio and the relationship between the maximum and minimum stress intensity factors (SIF) for the preferential propagation path show initially high values that decrease slowly in the case of tension loading and sharply in the case of bending loading.

(iv) The increase of the Paris exponent *m* reduces the crack aspect ratio for the preferential propagation path, increasing the ratio between the minimum and maximum SIFs.

Acknowledgements

The authors wish to acknowledge the financial support provided by the following Spanish Institutions: Ministry for Science and Technology (MICYT; Grant MAT2002-01831), Ministry for Education and Science (MEC; Grant BIA2005-08965), Ministry for Science and Innovation (MICINN; Grants BIA2008-06810 and BIA2011-27870) and *Junta de Castilla y León* (JCyL; Grants SA067A05, SA111A07 and SA039A08).

References

1. J.C. Newman Jr., and I.S. Raju, "Analyses of Surface Cracks in Finite Plates under Tension or Bending Loads" (NASA TP-1578, USA, 1979).

2. I.S. Raju, and J.C. Newman Jr., "Stress-Intensity Factors for a Wide Range of Semi-Elliptical Surface Cracks in Finite-Thickness Plates", **Engineering Fracture Mechanics**, 11 (1979), 817-829.

3. S.J. Holdbrook, and W.D. Dover, "The Stress Intensity Factor for a Deep Surface Crack in a Finite Plate", **Engineering Fracture Mechanics**, 12 (1979), 347-364.

4. M. Isida, H. Noguchi, and T. Yoshida, "Tension and Bending of Finite Thickness Plates with a **Semielliptical Surface Crack", International** Journal of Fracture, 26 (1984), 157-188.

5. M. Isida, H. Noguchi, and T. Yoshida, "Oblique Semi-Elliptical Surface Crack in Semi-Infinite Solid Subjected to Tension", **Engineering Fracture Mechanics**, 36 (1990), 889-892.

6. X.B. Lin, and R.A. Smith, "Finite Element Modelling of Fatigue Crack Growth of Surface Cracked Plates. Part I: The Numerical Technique", **Engineering Fracture Mechanics**, 63 (1999), 503-522.

7. Z. Wu, "The Shape of a Surface Crack in a Plate Based on a Given Stress Intensity Factor Distribution", International **Journal of Pressure Vessels and Piping**, 83 (2006), 168-180.

8. Y. Peng, L. Tong, X.-L. Zhao, and Z. Xiao, "Modified Stress Intensity Factor Equations for Semi-Elliptical Surface Cracks in Finite Thickness and Width Plates", **Procedia Engineering**, 14 (2011), 2601-2608.

9. O. Vosikovsky, and A. Rivard, "Growth of Surface Fatigue Cracks in a Steel Plate", **International Journal of Fatigue**, 3 (1981), 111-115.

10. A. Hosseini, and M.A. Mahmoud, "Evaluation of Stress Intensity Factor and Fatigue Growth of Surface Cracks in Tension Plates", **Engineering Fracture Mechanics**, 22 (1985), 957-974.

11. M.A. Mahmoud, and A. Hosseini, "Assessment of Stress Intensity Factor and Aspect Ratio Variability of Surface Cracks in Bending Plates", **Engineering Fracture Mechanics**, 24 (1986), 207-221.

12. X.B. Lin, and R.A. Smith, "Finite Element Modelling of Fatigue Crack Growth of Surface Cracked Plates. Part II: Crack Shape Change", **Engineering Fracture Mechanics**, 63 (1999), 523-540.

13. J.C. Newman Jr., and I.S. Raju, "An Empirical Stress-Intensity Factor Equation for the Surface Crack", **Engineering Fracture Mechanics**, 15 (1981), 185-192.

14. P.C. Paris, and F. Erdogan, "A Critical Analysis of Crack Propagation Laws", **Journal of Basic Engineering**, 85D (1963), 528-534.

15. X.B. Lin, and R.A. Smith, "Finite Element Modelling of Fatigue Crack Growth of Surface Cracked Plates. Part III: Stress Intensity Factor and Fatigue Crack Growth Life", **Engineering Fracture Mechanics**, 63 (1999), 541-556.

16. I.S. Putra, and J. Schijve, "Crack Opening Stress Measurements of Surface Cracks in 7075-T6 Aluminium Alloy Plate Specimen Through Electron Fractography", **Fatigue & Fracture of Engineering Materials & Structures**, 15 (1992), 323-338.

17. N.B. McFadyen, R. Bell, and O. Vosikovsky, "Fatigue Crack Growth of Semielliptical Surface Cracks", **International Journal of Fatigue**, 12 (1990), 43-50.

AUTHOR INDEX
Fatigue of Materials III: Advances and Emergences in Understanding

A

B

C

D

E

F

G

I

K

L

M

N

O

P

Q

R

S

T

V

Z

SUBJECT INDEX
Fatigue of Materials III: Advances and Emergences in Understanding

A

B

C

D

E

F

H

L

M

N

O

P

R

T

U

MIX
Papier aus verantwortungsvollen Quellen
Paper from responsible sources
FSC® C105338

If you have any concerns about our products,
you can contact us on
ProductSafety@springernature.com

In case Publisher is established outside the EU,
the EU authorized representative is:
Springer Nature Customer Service Center GmbH
Europaplatz 3, 69115 Heidelberg, Germany

Printed by Libri Plureos GmbH
in Hamburg, Germany